Complex Analysis: An Introduction

Dr Kevin Houston follows up his best-selling book *How to Think Like a Mathematician* with *Complex Analysis: An Introduction*. Complex Analysis is a central subject in mathematics with applications in engineering, physics, and even the study of prime numbers. It has been said that often the shortest route in the solution of a real problem is to take a shortcut through the complex numbers.

Unlike other texts this book gets quickly to the heart of Complex Analysis: the concept of complex contour integration. This means that students get much more practice in the fundamental concept than they normally would. The central method of proof – use of the Estimation Lemma – is emphasised throughout because students then have a unifying principle to help understand and remember those proofs.

The book contains all you will need for an introductory course in Complex Analysis and includes a short and sweet proof of Cauchy's Theorem – one which the majority of students can grasp not only the outline but the details as well.

The book contains copious examples and exercises tested on students arising from Dr. Houston's 20 years plus experience of teaching the subject.

Complex Analysis
An Introduction

KEVIN HOUSTON

Complex Analysis: An Introduction Version 1.0.

ISBN: 978-1-9997952-0-7

Published by x to the power of n.

www.xtothepowerofn.com

Cover background designed by Ydlabs / Freepik

This one is for Katie.

Contents

Preface

Mathematicians love to generalize, to make a statement cover more situations. For example, we start with Pythagoras' Theorem which relates the sides of a right angled triangle by $c^2 = a^2 + b^2$ where c is the length opposite the right angle. It is natural to ask what happens when the angle is not a right angle. Well, in that case we have the Cosine Rule: $c^2 = a^2 + b^2 - 2ab \cos C$ where C is the angle opposite the length c. This reduces to Pythagoras' Theorem when C is a right angle and hence is a generalization.

As mathematicians we can generalize our concept of number from real numbers to complex numbers. Pretending that the square root of -1 exists gives us more streamlined theorems – all quadratics have two roots (counting multiplicity). Furthermore, real problems with real solutions can be solved by taking a detour through the complex plane. For example, if the auxiliary equation of a 2nd order linear differential equation with constant coefficients has complex roots, then we know that the solution involves sines and cosines (rather than exponentials).

Real analysis can be crassly defined as the task of making rigorous the concepts of calculus, i.e., ensuring that the definitions of differentiation and integration make sense and their associated theorems are true. The aim of this book is to introduce **complex analysis**. This is a generalization of calculus from real functions to complex functions.

Calculus is the mathematical tool par excellence for solving real-world physical problems and we can probably justify such a generalization on these grounds. Nonetheless, we must take care that we are not merely generalizing for the sake of generalizing. Fortunately, victories for complex analysis quickly emerge and, ultimately, we see the appealing pattern that a detour through complex numbers can solve real problems with simplicity and efficiency.

The first of these victories is that differentiable complex functions give solutions to Laplace's equation, a differential equation related to many important physical problems, for example, fluid dynamics, electromagnetism, and heat flow.

Some of the results are charmingly quirky. For example, Corollary 18.4 states that for a real polynomial p of degree $n \geq 2$ with real distinct roots α_1, α_2, ..., α_n, we have

$$\sum_{j=1}^{n} \frac{1}{p'(\alpha_j)} = 0.$$

Who would have guessed that the gradients of the function at its roots were related in this way? It is not clear from real analysis that this is the case but it is remarkably simple to discover using complex analysis. In fact, one can prove it using only real methods but it does get very messy.

The central results are more than just curios. For example, in Chapter 19 we calculate real integrals that would be nearly impossible to find using real techniques. In Chapter 21 we show that $\sum_{n=1}^{\infty} 1/n^2 = \pi^2/6$ along with other results regarding summation of series. In both cases we have real problems that are hard to solve with real analysis methods but are simple with complex analysis methods.

Ultimately, we get probably the most powerful set of techniques in mathematics for solving problems. On the other hand, one of the most satisfying aspects of complex analysis is that the theorems are aesthetically pleasing. Cauchy's Integral Formula allows us to find the value of a complex differentiable function by integrating almost any collection of other values. And, in sharp contrast to real analysis, if a function is complex differentiable once, then it is infinitely complex differentiable. Furthermore, again in contrast to real analysis, the function is equal to its Taylor series. These theorems are in Chapter 13.

The theoretical part of complex analysis climaxes in this book with Cauchy's Residue Formula, which says that we can calculate certain complex integrals by summing what are called residues. These are surprisingly easy to calculate.

Complex analysis is a central part of mathematics and has much to recommend it. It is theoretical pleasing and can be applied directly to real-world problems. What more could we ask of a mathematical theory?

The approach in this book

In this book I have tried as much as possible to motivate concepts in natural and intuitive ways. For example, the question of how one defines the exponential, sine and cosine functions leads one to consider defining them as series of complex numbers and this naturally leads one to the develop the theory of complex series.

Nonetheless, the book takes an unconventional route through complex analysis. Traditionally differentiation precedes integration. This makes sense as ordinary real calculus is taught this way (though one could argue that as Archimedes more-or-less invented integration nearly 2000 years before Newton and Leibniz invented differentiation it may not make sense from a historical perspective). In this book the order is reversed, the notion of complex integration, called contour integration, precedes complex differentiation.

The first reason is that contour integration is central in complex analysis in a way that integration is not in real analysis and hence studying integration first helps signal this centrality. Furthermore, to put complex differentiation early in the course would fool students into thinking that complex analysis is merely a trivial extension of real analysis, whereas presenting contour integration shows students something new and exciting early on and, crucially, gives more time for the concept to sink in.

The proofs in the book are constructed to be as simple and as efficient as possible whilst never losing sight of understanding. I have spent many years searching for good proofs, adapting them and testing them on students to make sure they are understandable. In fact, I believe that there is only one proof that uses a fancy trick, Theorem 14.10, as I thought it was too good a trick to leave out. The others are optimized for understanding.

Students, a moment of your valuable time

My main piece of advice to students is that you learn mathematics by doing mathematics. You do not learn it by passively reading or listening. You learn it by taking care to notice the precise words in a definition or theorem and, most importantly, by doing the exercises. One of the best techniques for getting in to a subject is to do exercises involving calculation so look for those. They are an effective way into a subject.

Even when reading you should have a pen and paper near by to explore the definitions and theorems. For example, to create some examples and non-examples of a definition or to do the calculation that is skipped over in a proof. Many more techniques regarding learning can be found in my book *How to Think Like a Mathematician.*

Turning to complex analysis, my advice depends on how you view mathematics. If you like real world applications, then these are mainly at the end of the book and it requires perseverance to reach the payoff. You have to invest in an initially strange and difficult-to-follow topic. Fortunately, the subject gets easier as it progresses. The important concept to reach is that of residue, and the residues of functions are actually laughably easy to calculate, they generally involve nothing more than differentiating and plugging in values. And we then use them to solve with ease real integrals that would be frighteningly tough to find using the techniques of real analysis.

If you prefer mathematical theory to applications, then there is much here to interest you. However, the really exciting parts, such as Cauchy's Theorem, Integral Formula and Residue Theorem don't appear until midway through the book so one has to delay gratification until then. Nonetheless, they are well worth waiting for.

For many of my students a major stumbling block with exercises is not the complex analysis but a lack of fluency with the basics of complex numbers. A problem with complex arithmetic rather than complex analysis! For example, they are confused as to the behaviour of the modulus operator. (That $\sqrt{x^2}$ is not equal to x but is equal to $|x|$ often causes surprise in my class.) Hence, although the chances are that the material in Chapter 1 is not new to you make sure you know and understand it well.

One major difference between this book and others is that explanations of common student errors and misunderstandings are sprinkled through text and there is a whole chapter devoted to them. Hopefully, these will help you avoid making them.

And, lecturers, a moment too of your valuable time

I have never had a student who didn't understand complex differentiation almost immediately. Teaching integration first means the difficult definition, the concept of contour integration, is met early to give students more time to learn it. My approach is to define complex integration as early as possible and set some examples for students to tackle *in the lecture* (I use Exercises 5.12). This has the advantage of flagging up the importance of the topic and gives students the chance to immediately engage with the definition. Introducing it early also means that I can set exercises for them to do, hand in, have marked and returned. By following the complex integration with differentiation (including differentiability of complex series which logically speaking does not have to appear until after Cauchy's Theorem in preparation for Taylor's Theorem) means that the students have a chance to do these exercises and understand integration before they encounter Cauchy's Theorem. To define complex integration after differentiation and immediately follow it with Cauchy's Theorem means that the students do not have time to digest integration and so the theorem looks meaningless. It looks like the lecturer has miswritten the limits of integration – forgetting the b and putting a γ where the a should be!.

As you probably know, the Estimation Lemma (Theorem 7.5) is the main tool in complex analysis. I've tried to ensure that where it is used is clear. I even use it in the proof of the Fundamental Theorem of Algebra (Theorem 14.1) rather than deduce the theorem from Liouville's Theorem and the Max Modulus Principle as is usual. Just as for the definition of contour integral, I ensure that students tackle the Estimation Lemma by giving them time in a lecture to do Exercise 7.9.

The biggest problem in any complex analysis course is how to tackle the proof of Cauchy's Theorem. To do so rigorously after the statement means that a lot of time is spent on hard technical details which are never used again. To avoid this lecturers sometimes leave the proof to the end of the course and this means that the most difficult part is stated when the students, tired by a long course, are at their least receptive to technical ideas. (Needless to say that sometimes the proof is then omitted due to time constraints.)

The usual proofs of Cauchy's Theorem are not particularly illuminating. For example, the proof from Beardon that I used for many years does not give one much insight as to why the theorem is true. Even if students could follow it. (I doubt that 1% of students of the hundreds of students I have taught in the course truly engaged with it and grasped even the outline.)

My approach is to prove the theorem for a restricted class of contours - those made of a finite number of straight lines and arcs. One could argue that the resulting proof is still intuitive in places but I think that the compromises made are no different from those made in the Beardon proof. The proof does have numerous advantages:

1. It is enough for our purposes - all the applications we meet involve such contours.

2. It is short and so students are more likely to engage with it.

3. It follows one of the principles of the book - use the Estimation theorem as much as possible.

4. Due to the simplicity of the theorem it is easier to see why the theorem is true.

For those wishing to give a full proof, then see Appendix A and ensure that Exercise 11.12(ii) is covered.

Standing on the shoulders of giants

It is rare for a textbook to introduce revolutionary original material, particularly complex analysis, a central topic of mathematics first developed in the 19th century. Hence, the material in this book is expository though I hope I can claim a little originality for the route taken through the world of complex analysis.

This book is the culmination of many years work, either on courses I have taught many times at the University of Leeds or at courses I assisted on at the University of Warwick and University of Liverpool whilst a postgraduate and research assistant respectively. My notes for the courses at Leeds changed over the years as topics were rewritten to simplify proofs or improve exposition and have been tested on many students.

Nonetheless, I am directly indebted to two major influences for this book. The first is to my complex analysis teacher, Prof. John Ringrose, from whom I took the idea that teaching integration first was the best way to go. Second, the book owes a large debt to my colleagues at Leeds. My notes grew out of a set I inherited from Prof. Jonathan Partington who had in turn inherited his from others. The overall skeleton of the book, selection of material and some phrasing in theorems and definitions are derived from those notes. In some cases, for example the proof of Theorem 14.10 and Example 17.6(iii), are taken almost verbatim from them.

Over the years I picked up many ideas without necessarily noting the originator. However, I would like to record that the corollary of Theorem 18.3 was shown to me by David Chillingworth (in the context of gravitational lensing!).

Despite borrowing from others, needless to say, the responsibility for all errors are mine. If you spot any, then please get in touch via email. (I'm particularly concerned about errors in the chapter on common errors as it is very embarrassing to admonish others while making mistakes even if they are just typos.)

Acknowledgements

First, I would like to thank my students. They have been unwittingly experimented upon in my quest to find the best proof or explanation. Also, many have spotted typos and other mistakes. My thanks to all of them.

Thanks are due to my colleagues at Leeds, Jonathan Partington of course, but also Alan Slomson, Reg Allenby, John Wood, Martin Speight and Roger Bielawski for stimulating discussions about how to teach complex analysis. Special thanks to Katie Chicot for providing helpful lecture notes.

Kevin Houston
May 2017 Leeds, England
www.kevinhouston.net

CHAPTER 1

Complex Numbers

Complex numbers arise from imagining the consequences of the existence of the square root of -1. This flight of fancy is one of the most powerful theoretical tools invented. It is likely you will have already met complex numbers and seen them used to solve problems such as the description of simple harmonic motion in the study of ordinary differential equations. We shall see in this book how extending calculus to the domain of complex numbers allows us to solve numerous real problems such as the solution of other ordinary differential equations, evaluation of integrals, and summation of series. Furthermore, the subject is interesting and puzzling in its own way and we can derive pleasure from its study.

In this chapter we shall revise basic definitions of complex numbers. It is vital that you are entirely comfortable with this material and able to use it without thinking deeply. I see some students struggle in my complex analysis course because of their problems with complex 'arithmetic', that is, with concepts such as absolute value.

Complex numbers without the fantasy

Complex numbers are usually defined to be numbers of the form $x + iy$ where x and y are real numbers and $i = \sqrt{-1}$. We can manipulate these numbers through addition and multiplication in the usual way subject to the rule $i^2 = -1$.

Hence, complex numbers are the numbers we get when we pretend that the square roots of negative numbers exist. This is problematic as it is illogical. The square root of -1 clearly does not exist and so it appears we are building a mathematical system on a fantasy. To avoid this we can define complex numbers as pairs of real numbers and define addition and multiplication on that set of pairs in such a way that the resulting number system is equivalent to the system with the imaginary number $\sqrt{-1}$ in it.

Definition 1.1
The **complex number system**, denoted $\mathbb{C}$, is the set of ordered pairs of real numbers with the operations addition, denoted $+$, and multiplication, denoted $\times$, defined by

$$(a, b) + (c, d) = (a + c, b + d)$$
$$(a, b) \times (c, d) = (ac - bd, ad + bc)$$

for all (a, b) and (c, d) in $\mathbb{R}^2$. A **complex number** is an element of this number system.

It is straightforward but to tedious to check that these operations are commutative, associative and distributive. The point is that this definition requires no flights of imagination such as the existence of the square root of -1.

Now we need to show that this system corresponds to what we know already as the set complex numbers, i.e., numbers of the form $x + iy$. First, for all real numbers α we can define $\alpha(a, b) = (\alpha a, \alpha b)$. With this we can write any (a, b) as

$$(a, b) = a(1, 0) + b(0, 1).$$

We can make a correspondence of $a + bi$ with (a, b) by identifying $(0, 1)$ with the number 1 and $(0, 1)$ with i.

Consider the pair $(0, 1)$ and its square. According to the definition of multiplication we have

$$(0, 1) \times (0, 1) = (0 \times 0 - 1 \times 1, 0 \times 1 + 1 \times 0) = (-1, 0).$$

Under our correspondence this corresponds to $i^2 = -1$.

It is again straightforward and tedious to show that under this correspondence addition and multiplication defined for the pairs correspond to addition and multiplication of complex numbers in the form $x + iy$.

This means that the system of numbers defined using the plane $\mathbb{R}^2$ is equivalent to that of the complex numbers defined using the square root of -1. This equivalence dispels any feelings of unease about building mathematics based on objects which cannot logically exist.

Hence, complex numbers rest not on a belief of imaginary objects but on a belief of the plane.

Basic Definitions

Although one uses the plane to define complex numbers, in practice no one views complex numbers as multiplication and addition on the plane. We use square root of -1 form, confident that this is equivalent to a rigorously defined system. We call the form $x + iy$ the **Cartesian form** or **Cartesian representation** of complex numbers

It is traditional to denote complex numbers with the letters z and w and for z to be written in Cartesian form as $x + iy$ where x and y are real numbers. Therefore, let z be a complex number with $z = x + iy$ its Cartesian representation and $x, y \in \mathbb{R}$. The **real part of** z, denoted $\operatorname{Re}(z)$, is the number x. The **imaginary part of** z, denoted $\operatorname{Im}(z)$, is the number y. So one trivially has $z = \operatorname{Re}(z) + i\operatorname{Im}(z)$.

The **complex conjugate** of z, denoted $\bar{z}$, is the complex number $x - iy$.

Proposition 1.2

For all $z, w \in \mathbb{C}$, we have

(i) $\overline{zw} = \bar{z}\,\bar{w}$,

(ii) $\overline{z + w} = \bar{z} + \bar{w}$, and

(iii) $\bar{\bar{z}} = z$.

These are easily proved by converting to Cartesian representation. Furthermore, the conjugate is useful for converting quotients of complex numbers into Cartesian form:

$$\frac{1}{z} = \frac{1}{x + iy} = \frac{1}{x + iy}\frac{x - iy}{x - iy}$$
$$= \frac{x - iy}{x^2 - ixy + ixy - i^2y^2} = \frac{x - iy}{x^2 + y^2}.$$

The next definition is crucial in complex analysis. It allows us to convert complex number problems to real number problems.

Definition 1.3

The **modulus** of z, denoted $|z|$, is the non-negative number given by

$$|z| = \sqrt{x^2 + y^2}.$$

This is also called the **absolute value** of z and sometimes (though not often in this book) referred to as the **length** of z.

Proposition 1.4

For any complex numbers $z, w \in \mathbb{C}$ we have

(i) $|z|^2 = z\bar{z}$,

(ii) $|zw| = |z||w|$,

(iii) $\dfrac{1}{z} = \dfrac{\bar{z}}{|z|^2}$,

(iv) $|\mathrm{Re}(z)| \leq |z|$ and $|\mathrm{Im}(z)| \leq |z|$,

(v) $|z| = 0 \iff z = 0$,

(vi) $|\bar{z}| = |z|$.

Proof. (i) This is proved by a straightforward calculation.

(ii) To prove this we consider the square:

$$|zw|^2 = zw\overline{zw} = zw\bar{z}\,\bar{w} = z\bar{z}w\bar{w} = |z|^2|w|^2$$

and since $|z|$, $|w|$ and $|zw|$ are all non-negative we have $|zw| = |z||w|$.

(iii) Follows from (i)

$$\frac{1}{z} = \frac{1}{z}\frac{\bar{z}}{\bar{z}} = \frac{\bar{z}}{|z|^2}.$$

(iv) These are geometrically obvious when we consider an Argand diagram (defined in the next section). A rigorous proof is as follows:

$$|\mathrm{Re}(z)|^2 = x^2 \leq x^2 + y^2 = |z|^2$$

and since $|\mathrm{Re}(z)|$ and $|z|$ are non-negative the result follows. Similarly for $\mathrm{Im}(z)$.

(v) This follows simply from $|z| = \sqrt{x^2 + y^2}$.

(vi) The proof is a simple calculation. $\qquad\qquad\square$

Remarks 1.5

(i) In proofs of (ii) and (iv) there is a useful trick: use the square of the modulus and hence operate with numbers and their conjugates rather than messy square root signs. This will be used a few times in the book.

(ii) That $|z| = 0$ implies $z = 0$ is an important result and is used at numerous places in the book.

Common Error 1.6

The modulus of a complex number is always a non-negative *real number*. A very common error seen in homework and exam solutions is a calculation that produces an imaginary number. This occurs because the student squares iy rather than y. For example for $z = 3 + 4i$, rather than the correct $\sqrt{3^2 + 4^2}$ they calculate $\sqrt{3^2 + (4i)^2}$ and this gives $\sqrt{9 - 16} = \sqrt{-7} = \sqrt{7}i$. Clearly, this cannot be right as the modulus is real.

Argand diagrams, Polar and Exponential Forms

Since complex numbers are pairs of real numbers we can plot them in the plane. The resulting diagram is called an **Argand diagram**.

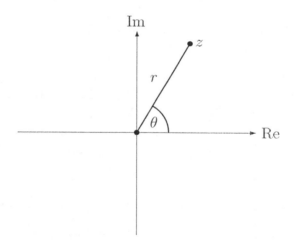

For each $z \neq 0$, we can measure the angle between the positive x-axis and the ray from the origin to z. This angle is denoted θ in the diagram above and is usually measured in radians. By taking anti-clockwise as positive we can define the **principal value of the argument**, denoted $\text{Arg}(z)$, to be the value of θ such that $-\pi < \theta \leq \pi$. (Note the strict inequality.) For $z = 0$ we do not define the argument.

From the diagram it is clear that if $z = x + iy$, r is the modulus of z and θ is the principal value of the argument, then

$$x = r\cos\theta \text{ and } y = r\sin\theta.$$

Hence, we can write

$$z = r\cos\theta + ir\sin\theta = r\left(\cos\theta + i\sin\theta\right).$$

The expression $r\left(\cos\theta + i\sin\theta\right)$ is called the **polar form** of z as it arises from the polar coordinates (r, θ).

We can see that $r = |z| = \sqrt{x^2 + y^2}$ but the calculation of θ from x and y is a little more subtle. We have, for $x \neq 0$,

$$\frac{y}{x} = \frac{r\sin\theta}{r\cos\theta} = \tan\theta.$$

It appears that we can use $\tan^{-1}$ to find $\mathrm{Arg}(z)$. In fact, in many texts the argument for $z = x + iy$ is *defined* to be $\theta = \tan^{-1}(y/x)$.

However, tan is a periodic function (i.e., $\tan(\theta + 2\pi) = \tan(\theta)$) so, strictly speaking, no inverse exists. Instead, since tan restricted to $(-\pi/2, \pi/2)$ has an inverse it is this we use for the definition of 'inverse' and we denote it by arctan. That is $\arctan = \tan^{-1}|_{(-\pi/2, \pi/2)}$. In fact one cannot just say $\theta = \mathrm{Arg}(x + iy) = \arctan(y/x)$.

This situation requires care, we cannot use arctan and $\tan^{-1}$ interchangeably, and the following error often occurs.

Common Error 1.7

Calculating θ using $\arctan(y/x)$ (or $\tan^{-1}(y/x)$) leads to problems for $x < 0$. For example, from an Argand diagram it is clear that

$$\mathrm{Arg}(1 + i) = \frac{\pi}{4} \quad \text{and} \quad \mathrm{Arg}(-1 - i) = -\frac{3\pi}{4}.$$

However,

$$\arctan(1/1) = \arctan(1) = \arctan(-1/-1).$$

That is, the principal arguments are different but the arctan formula gives the same number.

We can use the $\arctan(y/x)$ formula provided we take care to add π or subtract depending on the position of the number on the Argand diagram. For the record we have

$$\mathrm{Arg}(x + iy) = \begin{cases} \arctan(y/x), & x > 0, \\ \pi + \arctan(y/x), & x < 0, y \geq 0, \\ -\pi + \arctan(y/x), & x < 0, y < 0, \\ \pi/2, & x = 0, y > 0, \\ -\pi/2, & x = 0, y < 0, \\ \text{undefined}, & x = y = 0. \end{cases}$$

It is useful to extend our notion of argument to include the case where θ is any real number. That is, **an argument of** $z = x + iy$ is any θ such that

$$x = r\cos\theta \text{ and } y = r\sin\theta.$$

We shall denote any such θ as $\arg(z)$. To be truly rigorous then we should treat $\arg(z)$ as the set containing all such possible θ but this is rarely done by mathematicians as, since we see later, we have the useful and pleasing formula $\arg(zw) = \arg(z) + \arg(w)$.

We can now define the exponential of a complex number. First, one can *define* $e^{i\theta}$ for a real number θ as

$$e^{i\theta} = \cos\theta + i\sin\theta.$$

(This definition is perfectly usable and is likely to be the one you have met. However, in Chapter 2 we shall give a definition via a series.) Next, for $z = x + iy$ we define

$$e^z = e^x e^{iy}$$

and hence

$$e^z = e^x \cos y + ie^x \sin y.$$

Let z be any complex number, then we can write z in the form $re^{i\theta}$ for some real numbers $r \geq 0$ and θ. This is called **exponential form** of z. The number r is just the modulus of z and θ is an argument of z. Due to the periodic nature of sin and cos we see that $re^{i\theta + 2k\pi} = re^{i\theta}$ for any $k \in \mathbb{Z}$.

Exponential notation is particularly useful when multiplying complex numbers. If we have $z_1 = x_1 + iy_1$ and $z_2 = x_2 + iy_2$, then $z_1 z_2 = (x_1 + iy_1)(x_2 + iy_2)$ involves multiplying out a bracket. However, in exponential form $z_1 = r_1 e^{i\theta_1}$ and $z_2 = r_2 e^{i\theta_2}$ and then, $z_1 z_2$ is simply $r_1 r_2 e^{i(\theta_2 + \theta_2)}$. That is we need only multiply two real numbers and add two real numbers. Note that we have $\arg(z_1 z_2) = \arg(z_1) + \arg(z_1)$. This does not hold for Arg.

To prove the preceding rigorously involves proving some further facts following from the definition of exponential given above. As the exponential of a complex number will be defined later in a different way we will neither state nor prove these.

To conclude this section we note the following famous theorem.

Proposition 1.8 (De Moivre's Theorem)
For all $\theta \in \mathbb{R}$ and $n \in \mathbb{Z}$ we have

$$(\cos\theta + i\sin\theta)^n = \cos n\theta + i\sin n\theta.$$

The proof follows by induction using products written in exponential notation.

Triangle Inequality

The next error is one of the most common.

Common Error 1.9

In contrast to the real numbers we cannot order the complex numbers. That is, although it is fine to write $2 \leq 5$ or $4 < 7$ we cannot write $3 + i \leq 4 + 2i$ or $2 - i > 7$ for example.

Though we cannot directly compare complex numbers the modulus function allows us to produce inequalities involving them. The most useful expression of this is the triangle inequality.

Proposition 1.10 (Triangle Inequality)

Let z and w be complex numbers. Then

$$\big||z| - |w|\big| \leq |z + w| \leq |z| + |w|.$$

Proof. We have

$$\begin{aligned}
|z + w|^2 &= (z + w)\overline{(z + w)} \\
&= z\overline{z} + z\overline{w} + w\overline{z} + w\overline{w} \\
&= |z|^2 + 2\mathrm{Re}(z\overline{w}) + |w|^2 \\
&\leq |z|^2 + 2|z\overline{w}| + |w|^2 \\
&= |z|^2 + 2|z||w| + |w|^2 \\
&= (|z| + |w|)^2.
\end{aligned}$$

Since the terms inside the squares on both sides of the inequality are non-negative real numbers we deduce that the second inequality holds.

Now for the first inequality. Using the

$$|z| = |z + w - w| \leq |z + w| + |-w|$$

and so

$$|z| - |w| \leq |z + w|.$$

From this we can see that

$$|w| - |z| \leq |w + z| = |z + w|$$

and combining these two inequalities involving $|z - w|$ we deduce $|z + w| \geq \big||z| - |w|\big|$. $\square$

The inequality $|z + w| \leq |z| + |w|$ is usually called the **triangle inequality**. The result and the reason for its name can easily be seen in the following Argand diagram. The inequality $\big||z| - |w|\big| \leq |z + w|$ is sometimes called the **reverse triangle inequality**.

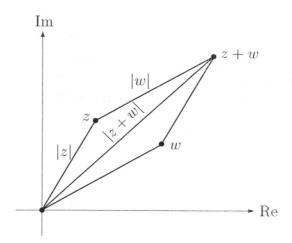

Roots of unity and similar equations

Definition 1.11

The solutions of $z^n = 1$ are called the nth **roots of unity**.

Example 1.12

The 3rd roots of unity, also called the **cube roots of unity**, are

$$1, \frac{-1 + i\sqrt{3}}{2} \text{ and } \frac{-1 - i\sqrt{3}}{2}.$$

This can be seen by simple but tedious calculation or using the following method.

Consider the equation $z^n = a$. By taking the modulus we get

$$|z^n| = |a|$$
$$|z|^n = |a|$$
$$|z| = \sqrt[n]{|a|}.$$

Next, for the argument

$$\arg(z^n) = \arg(a) + 2k\pi, \text{ for all } k \in \mathbb{Z},$$
$$n\arg(z) = \arg(a) + 2k\pi$$
$$\arg(z) = \frac{\arg(a) + 2k\pi}{n}$$

Hence,
$$z = \sqrt[n]{|a|}e^{i(\mathrm{Arg}(a)+2k\pi)/n} \text{ for all } k \in \mathbb{Z}.$$

In the case of roots of unity, i.e., $z^n = 1$, then $|a| = 1$ and $\mathrm{Arg}(a) = 0$. As angles differing by multiples of 2π are considered the same we see that we only get distinct values of $\mathrm{Arg}(z)$ for $0 \leq k < n$, that is we have, the following possible values for $\mathrm{Arg}(z)$:
$$0, \frac{2\pi}{n}, \frac{4\pi}{n}, \frac{6\pi}{n}, \ldots, \frac{2(n-1)\pi}{n}.$$

Therefore the roots of unity are
$$1, e^{2\pi i/n}, e^{4\pi i/n}, e^{6\pi i/n}, \ldots, e^{2(n-1)\pi/n}.$$

That is,
$$e^{2k\pi i/n} \text{ for } 0 \leq k < n \text{ with } k \in \mathbb{Z}.$$

Restricting further to $n = 3$ we get the following cube roots of unity
$$1, \quad e^{2\pi i/3} = \frac{-1 + i\sqrt{3}}{2} \quad \text{and} \quad e^{4\pi i/3} = \frac{-1 - i\sqrt{3}}{2}.$$

Solutions versus functions. Square roots

Common Error 1.13

Mathematicians are often sloppy about the meaning of square root. In general when asked for the square root of 64 many answer ± 8. Technically this is wrong. To see why we need to distinguish between a square root as the solution of an equation and as the value of a function.

The equation $x^2 = 64$ has the solutions 8 and -8. The function $\sqrt{}$ is such that $\sqrt{64} = 8$. The point is that the square root *function* can only have one value because functions are, by definition, single valued. In practice the function definition is the one implicitly used. We can see this since when solving the equation $ax^2 + bx + c = 0$ we write $x = \dfrac{-b \pm \sqrt{b^2 - 4ac}}{2a}$ with the $\pm$ indicating two solutions. If $\sqrt{}$ produced two solutions, then we need write only $x = \dfrac{-b + \sqrt{b^2 - 4ac}}{2a}$.

With this distinction in mind we can give a rigorous definition of the square root function on the set of non-negative real numbers: $\sqrt{a}$ for $a \geq 0$ is the positive solution to the equation $x^2 = a$.

The sloppiness in distinguishing between functions and solutions of equations leads to problems in understanding square roots for complex numbers. We can define the **square roots of the complex number** a to be the pair of solutions $z^2 = a$.

However, if we wish to define a **complex square root function**, then we have to take care. Let z be a complex number and let $z = re^{i\theta}$ be its exponential representation with $r \geq 0$ and $-\pi < \theta \leq \pi$, then

$$\sqrt{z} = \sqrt{r}e^{i\theta/2}.$$

It is straightforward to see that $(\sqrt{z})^2 = z$ and that when z is real this gives the familiar square root function.

A significant problem with this definition is that the function is not continuous along the negative axis. (The negative axis is called a **branch locus**.) This can be seen in an Argand diagram (on the next page) and is explained by the square root halving the argument. (Squaring a number doubles the argument so a square root function should halve it.) Consider the number -1, the square root of this is i. But a number close to -1 just below it will be taken to a number close to $-i$.

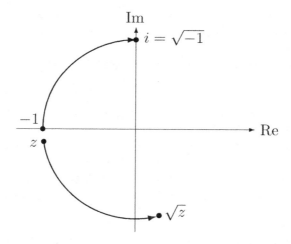

An important consequence of this is that, in sharp contrast to the real case, $\sqrt{zw} \neq \sqrt{z}\sqrt{w}$ for general z and w. If the equality held, then we could produce the following 'proof' of $-1 = 1$:

$$-1 = i^2 = \sqrt{-1}\sqrt{-1} = \sqrt{-1 \times -1} = \sqrt{1} = 1.$$

The third equality is where the error occurs; it uses $\sqrt{zw} = \sqrt{z}\sqrt{w}$.

Other definitions of square root lead to similar problems. One solution is to broaden our definition of function to the concept of multi-valued functions. Seeking to solve the problem also leads to the interesting concept of Riemann surface which we do not have space to investigate in this book.

Exercises

Exercises 1.14

(i) Find the modulus and principal argument for the following complex numbers and write in the form $a + bi$:

   (a) i (b) -1 (c) $-i$ (d) $2(\sqrt{3} - i)$

   (e) $(2 + i)$,   (f) $(2 + i)^7$ (g) $-\sqrt{3} - i$ (h) $4(1 - \sqrt{3}i)$

   (i) $\dfrac{7 + 3i}{2 + 5i}$ (j) $(1 - 2i)^{15}$ (k) $(1 - i)^{-7}$

(ii) Show that $\operatorname{Re}(iz) = -\operatorname{Im}(z)$ and $\operatorname{Im}(iz) = \operatorname{Re}(z)$ for all $z \in \mathbb{C}$.

(iii) Prove that

$$(a) \quad \operatorname{Re}(z) = \frac{z + \overline{z}}{2}, \qquad (b) \quad \operatorname{Im}(z) = \frac{z - \overline{z}}{2i}.$$

(iv) Show that $\operatorname{Arg}(\overline{z}) = -\operatorname{Arg}(z)$.

(v) Prove that $|z| \leq |\operatorname{Re}(z)| + |\operatorname{Im}(z)| \leq \sqrt{2}|z|$.

(vi) Show that
$$|z_1 + z_2 + \cdots + z_n| \leq |z_1| + |z_2| + \cdots + |z_n|.$$

(vii) Prove the Parallelogram Law

$$|z + w|^2 + |z - w|^2 = 2\left(|z|^2 + |w|^2\right)$$

for all $z, w \in \mathbb{C}$.

(viii) Prove that the Binomial Theorem holds for complex numbers.

(ix) Let p be a polynomial with real coefficients. Prove that $p(\overline{z}) = \overline{p(z)}$ and that if α is a root of p, then $\overline{\alpha}$ is too.

(x) Let $z^2 = a + bi$. Show that

$$z = \pm\left(\sqrt{\frac{a + \sqrt{a^2 + b^2}}{2}} + (\operatorname{sign} b)i\sqrt{\frac{-a + \sqrt{a^2 + b^2}}{2}}\right)$$

where $\operatorname{sign} b = 1$ if $b > 0$ and $\operatorname{sign} b = -1$ if $b < 0$. If $b = 0$, then $\operatorname{sign} b$ is irrelevant.

(xi) Find the solutions to the following equations

(a) $z^2 = 1 - i$ (b) $z^2 = 12 + 5i$ (c) $z^3 = \sqrt{3} - i$ (d) $z^4 = -i$.

(xii) Simplify

(a) $\sqrt{12 + 5i}$ (b) $\sqrt{1 + i}$ (c) $\sqrt{\sqrt{-i}}$ (d) $\sqrt{\sqrt{3 + 4i}}$.

(xiii) Sketch the following sets on an Argand diagram:

(a) $\{e^{it} + 2 \mid -\pi/2 \le t \le \pi\}$,
(b) $\{z \mid \pi/3 < \text{Arg}(z) < 2\pi/3 \text{ and } 1 < |z| < 2\}$,
(c) $\{(t - it)^3 \mid 0 \le t \le 1\}$,
(d) $\{te^{it} \mid 0 \le t \le \pi\}$.

(xiv) On an Argand diagram draw the image of z^2 for the set $\{z \in \mathbb{C} \mid \text{Re}(z) \ge 0, \text{Im}(z) \le 0\}$.

(xv) On an Argand diagram draw the image of the complex square root function applied to the set $\{z \in \mathbb{C} \mid 1 \le \text{Re}(z) \le 3\}$.

(xvi) Let z and w be complex numbers.

(a) Show that if $|z - w| \le |w|/2$, then $|z| \ge |w|/2$.
(b) Generalize the above to a statement where the hypothesis is $|z - w| \le |w|/c$ where $c \in \mathbb{R}$.
(c) Is it possible to generalize to a statement with the same hypothesis as (b) but with $c \in \mathbb{C}$ rather than $\mathbb{R}$? Give reasons.

(xvii) Define the nth root function $\sqrt[n]{} : \mathbb{C} \to \mathbb{C}$ by

$$\sqrt[n]{z} = \sqrt[n]{r}e^{i\theta/n} \text{ for } z = re^{i\theta}, \quad r \ge 0, \quad -\pi < \theta \le 0.$$

Calculate

(a) $\sqrt[4]{i}$ (b) $\sqrt[4]{-i}$ (c) $\sqrt[3]{2 + 3i}$ (c) $\sqrt[5]{242 + \sqrt{485i}}$.

(xviii) Using $e^{i\theta} = \cos\theta + i\sin\theta$ show that the sine and cosine addition formulae hold for real numbers. That is, for all $A, B \in \mathbb{R}$,

$$\sin(A \pm B) = \sin A \cos B \pm \cos A \sin B,$$
$$\cos(A \pm B) = \cos A \cos B \mp \sin A \sin B.$$

(xix) Prove Lagrange's Identity:

$$\left| \sum_{k=1}^{n} z_k w_k \right|^2 = \sum_{k=1}^{n} |z_k|^2 \sum_{k=1}^{n} |w_k|^2 - \sum_{1 \leq j < k \leq n} |z_j \overline{w}_k - z_k \overline{w}_j|^2$$

Hence, deduce the Cauchy-Schwartz inequality

$$\left| \sum_{k=1}^{n} z_k w_k \right|^2 = \sum_{k=1}^{n} |z_k|^2 \sum_{k=1}^{n} |w_k|^2$$

(xx) Show that

$$\frac{1 - z^n}{1 - z}$$

can be written as a polynomial. Use it to show that the sum of the nth roots of unity is 0.

Summary

- [ ] Complex number can be defined without resorting to numbers which do not exist. This is done by defining a special multiplication on the plane.

- [ ] Cartesian form: $z = x + iy$, where $x, y \in \mathbb{R}$.

- [ ] Polar form: $z = r(\cos\theta + i\sin\theta)$, $r \geq 0$, $\theta \in \mathbb{R}$.

- [ ] Exponential form: $z = re^{i\theta}$, $r \geq 0$, $\theta \in \mathbb{R}$.

- [ ] Triangle inequality: For all $z, w \in \mathbb{C}$,

$$\big| |z| - |w| \big| \leq |z + w| \leq |z| + |w|.$$

- [ ] nth roots of unity:

$$e^{2k\pi i/n} \text{ for } 0 \leq k < n \text{ with } k \in \mathbb{Z}.$$

- [ ] For $z = re^{i\theta}$ with $r \geq 0$ and $-\pi < \theta \leq \pi$, the complex square root function is

$$\sqrt{z} = \sqrt{r}e^{i\theta/2}.$$

CHAPTER 2

Complex Functions

In real analysis we study functions on open and closed intervals of $\mathbb{R}$, that is sets of the form (a, b) and $[a, b]$ respectively. We need to produce analogous definitions for functions on the set of complex numbers. To this end we shall first generalize the notion of open interval to an open set in the complex plane.

Furthermore, due to their importance in real function theory, we would like complex versions of the sine, cosine and exponential functions. We shall define these using complex power series and hence we need to define the notions of convergence for complex sequences and series. We shall also see how to use tools from real analysis, such as the ratio test.

We start by defining the analogue of open intervals.

Definition 2.1
An **open disc of radius** r **centred at** z is the set $\{w \in \mathbb{C} : |w - z| < r\}$ where r is a positive real number. An **open set** is a subset D of $\mathbb{C}$ such that for each point z in D there exists an r such that the open disc of radius r centred at z is contained in D.

Note that an open disc does not contain the circle on its boundary.

Examples 2.2
  (i) The set $D = \mathbb{C}$ is an open set. For $z \in \mathbb{C}$ any $r > 0$ will do for an open disc of radius r centred at z that is contained in $\mathbb{C}$.

  (ii) The set $D = \mathbb{C}\backslash\{0\}$ is an open set. For $z \in D$ let $r = \frac{1}{2}|z|$. Then the open disc of radius r centred at z is contained in D.

(iii) Let D be an open disc. Then $D = \{z : |z - a| < R\}$ for some $R > 0$ and $a \in \mathbb{C}$. For $z \in D$, let $r = \frac{1}{2}(R - |z - a|)$. This gives an open disc that is contained in D. Hence an open disc is an open set.

(iv) The set of real numbers $\mathbb{R}$ considered as a subset of $\mathbb{C}$ is not an open set. Consider any real number, then any open disc must contain some complex numbers, i.e., the disc does not lie wholly in the real numbers.

In real analysis we restricted our sets to intervals, i.e., we did not have $(a, b) \cup (c, d)$ where $b < d$ for example. For complex analysis we introduce a notion analogous to these sets.

Definition 2.3

A **domain** is a non-empty subset D of $\mathbb{C}$ such that D is an open set and cannot be written as $D = D_1 \cup D_2$ where D_1 and D_2 are non-empty disjoint open sets.

Example 2.4

The set $D = \{z \in \mathbb{C} \,|\, \mathrm{Re}(z)\mathrm{Im}(z) > 0\}$ is not a domain. We can take $D_1 = \{z \in \mathbb{C} \,|\, \mathrm{Re}(z) > 0, \; \mathrm{Im}(z) > 0\}$ and $D_2 = \{z \in \mathbb{C} \,|\, \mathrm{Re}(z) < 0, \; \mathrm{Im}(z) < 0\}$. Then $D = D_1 \cup D_2$ but D_1 and D_2 are non-empty, disjoint and open. The set D_1 is the upper right quadrant of the plane and D_2 is the lower left quadrant, as shown below.

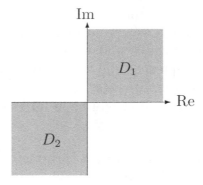

It can be shown that open discs are domains. We can now define the basic object of study.

Definition 2.5

Let D be an open set in $\mathbb{C}$. A **complex function**, denoted $f : D \to \mathbb{C}$, is a map which assigns to each z in D an element of $\mathbb{C}$, this value is denoted $f(z)$.

Note that for a complex function the truly important point is that the set upon which it is defined is an open subset of the complex numbers, not that the values are complex numbers.

Common Error 2.6

Note that f is the *function* and $f(z)$ is the *value* of the function at z. Technically, it is wrong to say $f(z)$ is a function, but sometimes people do.

Examples 2.7

The following are complex functions.

(i) Let $f(z) = z^2$ for all $z \in \mathbb{C}$.

(ii) Let $f(z) = |z|$ for all $z \in \mathbb{C}$. Note that here we have a complex function for which every value is real.

(iii) Let $f(z) = 3z^4 - (5 - 2i)z^2 + z - 7$ for all $z \in \mathbb{C}$. In fact, all complex polynomials on an open set are complex functions.

(iv) Let $f(z) = 1/z$ for all $z \in \mathbb{C}\backslash\{0\}$. This is a function that cannot be extended to all of $\mathbb{C}$.

Remark 2.8

Functions such as $\sin x$ for x real are not complex functions since the real line in $\mathbb{C}$ is not open. Later we see how to extend the concept of sine, cosine and exponential so that they are complex functions on the whole of the complex plane.

Obviously, if f and g are complex functions, then $f + g$, $f - g$, and fg are functions given by $(f+g)(z) = f(z)+g(z)$, $(f-g)(z) = f(z)-g(z)$, and $(fg)(z) = f(z)g(z)$, respectively. We can also define $(f/g)(z) = f(z)/g(z)$ provided that $g(z) \neq 0$ on D. Thus we can build up lots of new functions with these elementary operations.

The aim of complex analysis

We shall study complex functions and can now pose some questions. Can we define differentiation? Can we integrate? Which theorems from the analysis of real functions can be extended to complex functions? For example, is there a version of the mean value theorem? Complex analysis is the answer to these questions. The theory will be built upon real analysis but we will discover that it is easier than real analysis. For example, if a complex function is differentiable (in a sense to be defined later), then its derivative is also differentiable. This is not true for real functions. There exist differentiable functions f such that f' is not differentiable. (See Example 13.8 on page 136.)

Real and imaginary parts of functions

We will use u and v to denote the real and imaginary parts of a complex function $f : D \to \mathbb{C}$. That is, for $z = x + iy$ we can write

$$f(z) = f(x + iy) = u(x, y) + iv(x, y).$$

Note that u and v are functions of *two* real variables, x and y, i.e., $u : D \to \mathbb{R}$ and $v : D \to \mathbb{R}$.

Examples 2.9

(i) Let $f(z) = z^2$. Then, $f(x + iy) = (x + iy)^2 = x^2 - y^2 + 2ixy$. So, $u(x, y) = x^2 - y^2$ and $v(x, y) = 2xy$.

(ii) Let $f(z) = |z|$. Then, $f(x + iy) = \sqrt{x^2 + y^2}$. So, $u(x, y) = \sqrt{x^2 + y^2}$ and $v(x, y) = 0$.

Exercises 2.10

Find u and v for the following:

(i) $f(z) = 1/z$ for $z \in \mathbb{C} \backslash \{0\}$.

(ii) $f(z) = z^3$.

In the study of real functions we can graphically represent a function by drawing a graph. We have an axis for the variable and an axis for the value, and so we can draw the graph on a piece of paper.

A complex function requires two axes for the two variables, x and y, and two axes for the values of u and v. This means drawing a graph requires 4-dimensional space rather than the more convenient piece of paper.

For this reason the diagrams we draw to represent complex functions tend to be rather schematic as in the following diagram.

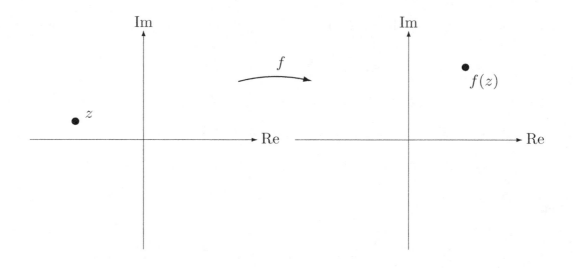

Defining e^z, $\cos z$ and $\sin z$

To help motivate our study we first define some elementary complex functions to play with. An obvious question is how shall we define e^z, $\cos z$ and $\sin z$? We require that their definition should coincide with the real version when z is a real number and we would like them to have properties similar to their real versions, e.g., $\sin^2 z + \cos^2 z = 1$ for all $z \in \mathbb{C}$ would be good.

However, real sine and cosine are initially defined using angles and it is not obvious how one should define a complex angle. The exponential is defined using differential calculus and we have not yet defined differentiation of complex functions.

Nonetheless, we know that the three real functions can be described using power series, e.g.,

$$\sin x = x - \frac{x^3}{3!} + \frac{x^5}{5!} - \cdots = \sum_{n=0}^{\infty} (-1)^n \frac{x^{2n+1}}{(2n+1)!} \text{ for all } x \in \mathbb{R}.$$

Thus, for $z \in \mathbb{C}$, we shall *define* the exponential, sine and cosine of z as follows:

$$e^z = \sum_{n=0}^{\infty} \frac{z^n}{n!},$$

$$\sin z = \sum_{n=0}^{\infty} (-1)^n \frac{z^{2n+1}}{(2n+1)!},$$

$$\cos z = \sum_{n=0}^{\infty} (-1)^n \frac{z^{2n}}{(2n)!}.$$

Thus,

$$e^{3+2i} = \sum_{n=0}^{\infty} \frac{(3+2i)^n}{n!} = 1 + (3+2i) + \frac{(3+2i)^2}{2!} + \frac{(3+2i)^3}{3!} + \cdots$$

These definitions obviously satisfy the requirement that they coincide with the definitions we know and love for the real versions, but how can we be sure that the three series converge? That is, does the series e^{3+2i} give a complex number?

To answer this we will have to study complex series. As the theory of *real series* was built on the theory of *real sequences* we had better start with complex sequences.

Remark 2.11

We can define the tangent function using a series as well but it is considerably simpler to define it by $\tan z = \dfrac{\sin z}{\cos z}$. This of course requires us to take care when the denominator is zero. See Exercise 3.16(vi).

Complex Sequences

The definition of convergence of a complex sequence is essentially the same as that for convergence of a real sequence.

Definition 2.12

A sequence of complex numbers z_n **converges** to $z \in \mathbb{C}$, if for all $\varepsilon > 0$ there exists a natural number N such that $|z_n - z| < \varepsilon$ for all $n \geq N$.

We write $z_n \to z$ or $\lim_{n \to \infty} z_n = z$ and call z the **limit of the sequence**.

Example 2.13

The sequence $z_n = \left(\dfrac{4 - 3i}{7} \right)^n$ converges to zero.

We have

$$|z_n - 0| = |z_n| = \left| \left(\frac{4 - 3i}{7} \right)^n \right| = \left| \frac{4 - 3i}{7} \right|^n = \left(\sqrt{\frac{25}{49}} \right)^n = \left(\frac{5}{7} \right)^n.$$

So

$$
\begin{aligned}
|z_n - 0| < \varepsilon \quad &\Longleftrightarrow \quad (5/7)^n < \varepsilon \\
&\Longleftrightarrow \quad n \log(5/7) < \log \varepsilon \\
&\Longleftrightarrow \quad n > \frac{\log \varepsilon}{\log(5/7)}.
\end{aligned}
$$

So, given any ε we can choose N to be greater than $\log \varepsilon / \log(5/7)$. Thus the sequence converges to zero.

Exercises 2.14

(i) Show that the sequence $z_n = z^n$ converges if and only if $|z| < 1$.

(ii) Consider the real sequence $a_n = |z_n - z|$. Show that $z_n \to z$ if and only if $a_n \to 0$.

Remark 2.15

The beauty of (ii) in the exercises above is that we are saying something about a complex sequence using real analysis. This is a good example of the paradigm we will use: Apply results from real analysis to produce results in complex analysis. In the above case we took the modulus to produce a real number. In other situations, like the next proposition, we will take real and imaginary parts.

Let's apply the paradigm. The next proposition shows that a sequence converges if and only its real and imaginary parts do.

Proposition 2.16

Suppose that $z_n = x_n + iy_n$ where x_n and y_n are real sequences, and suppose that $z = x + iy$ where x and y are real. Then

$$z_n \to z \iff x_n \to x \text{ and } y_n \to y.$$

Proof. We write $z_n - z = (x_n - x) + i(y_n - y)$.

[$\Rightarrow$] We have

$$0 \le |x_n - x| = |\mathrm{Re}(z_n - z)| \le |z_n - z|.$$

So if $z_n \to z$, then $x_n \to x$ by the squeeze rule and Exercise 2.14. Similarly, $y_n \to y$.

[⇐] We have

$$0 \le |z_n - z| = |(x_n - x) + i(y_n - y)| \le |x_n - x| + |y_n - y|$$

by the triangle inequality. Now, if $x_n \to x$ and $y_n \to y$, then, again by the squeeze rule and Exercise 2.14, $z_n \to z$.

□

Example 2.17
For the sequence $z_n = \dfrac{n^2 + in^3}{n^3 + 1}$ we have

$$\frac{n^2 + in^3}{n^3 + 1} = \frac{n^2}{n^3 + 1} + i\frac{n^3}{n^3 + 1} \to 0 + i.1 = i.$$

How to think like a mathematician 2.18
It is better to avoid using the definition of convergence to prove that a sequence converges.

Exercises 2.19
(i) Which of the following sequences converge(s)?

$$\frac{(n + 1)^5}{n^5 i} \qquad \text{and} \qquad \left(\frac{5 - 12i}{6}\right)^n.$$

Complex Series

Now that we have defined convergence of complex sequences we can define convergence of complex series.

Definition 2.20
Let z_n be a sequence of complex numbers. The complex series $\displaystyle\sum_{k=0}^{\infty} z_k$ **converges** if the sequence s_n formed by its partial sums $s_n = \displaystyle\sum_{k=0}^{n} z_k$ converges.

That is, the following sequence converges

$$\begin{aligned}
s_0 &= z_0 \\
s_1 &= z_0 + z_1 \\
s_2 &= z_0 + z_1 + z_2 \\
s_3 &= z_0 + z_1 + z_2 + z_3 \\
&\;\vdots
\end{aligned}$$

Let's apply the paradigm and give a result on complex series using real series.

Proposition 2.21

Let $z_k = x_k + iy_k$ where x_k and y_k are real for all k. Then,

$$\sum_{k=0}^{\infty} z_k \text{ converges } \iff \sum_{k=0}^{\infty} x_k \text{ and } \sum_{k=0}^{\infty} y_k \text{ converge.}$$

If either holds, then

$$\sum_{k=0}^{\infty} z_k = \sum_{k=0}^{\infty} x_k + i \sum_{k=0}^{\infty} y_k.$$

Proof. Let $a_n = \sum_{k=0}^{n} x_k$, $b_n = \sum_{k=0}^{n} y_k$, and $s_n = \sum_{k=0}^{n} z_k$. Then apply Proposition 2.16 to $s_n = a_n + ib_n$. The second part of the statement comes from equating real and imaginary parts. $\square$

Example 2.22

The complex series $\sum_{k=0}^{\infty} \dfrac{(-1)^k i}{k!}$ converges. Let $x_k = 0$ and $y_k = \dfrac{(-1)^k}{k!}$. Then $\sum x_k = 0$, obviously, and $\sum \dfrac{(-1)^k}{k!} = e^{-1}$.

Thus $\sum_{k=0}^{\infty} \dfrac{(-1)^k i}{k!}$ converges to $\dfrac{i}{e}$.

We have many great ways to decide whether a real series is convergent, for example, the divergence test, the ratio test and the integral test. Can we use these real tests in complex analysis? The next theorem says we can, but first we need a definition.

Definition 2.23

We say $\sum_{k=0}^{\infty} z_k$ is **absolutely convergent** or **converges absolutely** if the real series $\sum_{k=0}^{\infty} |z_k|$ converges.

This definition is the same as that in real analysis, it has merely been extended to complex numbers in a natural way.

Theorem 2.24

If the real series $\sum_{k=0}^{\infty} |z_k|$ converges, then $\sum_{k=0}^{\infty} z_k$ converges.

In other words,

absolute convergence implies convergence.

This is a powerful tool. The assumption is about *real* series (we know lots about these!) and gives a conclusion about *complex* series. Thus, given a complex series we can apply the ratio test, comparison test, etc, to a real series and say something about the complex series.

Proof. Let $z_k = x_k + iy_k$, with x_k and y_k real. Then $\sum_{k=0}^{\infty} |z_k|$ convergent implies that $\sum_{k=0}^{\infty} |x_k|$ converges (because $0 \leq |x_k| = |\text{Re}(z_k)| \leq |z_k|$ and we can apply the comparison test). So the real series $\sum_{k=0}^{\infty} x_k$ converges absolutely and this implies that $\sum_{k=0}^{\infty} x_k$ converges. Similarly, the series $\sum_{k=0}^{\infty} y_k$ converges.

Then, $\sum_{k=0}^{\infty} z_k = \sum_{k=0}^{\infty} x_k + i \sum_{k=0}^{\infty} y_k$ by Proposition 2.21. $\qquad\square$

How to think like a mathematician 2.25
To show that a series converges, show it converges absolutely.

Remark 2.26
Note that the converse to Theorem 2.24 is *not* true. We already know this from the real case. For example, $\displaystyle\sum_{k=0}^{\infty} \frac{(-1)^k}{k}$ converges but $\displaystyle\sum_{k=0}^{\infty} \left| \frac{(-1)^k}{k} \right| = \sum_{k=0}^{\infty} \frac{1}{k}$ diverges.

We now prove an infinite version of the triangle inequality.

Lemma 2.27
Suppose that $\sum_{k=0}^{\infty} z_k$ converges absolutely. Then

$$\left| \sum_{k=0}^{\infty} z_k \right| \leq \sum_{k=0}^{\infty} |z_k|.$$

Proof. For $n \geq 1$,

$$\left| \sum_{k=0}^{\infty} z_k \right| = \left| \sum_{k=0}^{\infty} z_k - \sum_{k=0}^{n} z_k + \sum_{k=0}^{n} z_k \right|$$

$$\leq \left| \sum_{k=0}^{\infty} z_k - \sum_{k=0}^{n} z_k \right| + \left| \sum_{k=0}^{n} z_k \right|$$

$$\leq \left| \sum_{k=0}^{\infty} z_k - \sum_{k=0}^{n} z_k \right| + \sum_{k=0}^{n} |z_k|.$$

As $n \to \infty$ then $\left| \sum_{k=0}^{\infty} z_k - \sum_{k=0}^{n} z_k \right| \to 0$, hence the result. $\qquad\square$

Another useful theorem is concerned with the product of series.

Theorem 2.28

Suppose that the series $\sum\limits_{n=0}^{\infty} a_n$ and $\sum\limits_{n=0}^{\infty} b_n$ are convergent such that at least one is absolutely convergent. Let $c_n = \sum\limits_{k=0}^{n} a_k b_{n-k}$. Then

$$\sum_{n=0}^{\infty} c_n = \left(\sum_{n=0}^{\infty} a_n \right) \left(\sum_{n=0}^{\infty} b_n \right).$$

The proof is essentially identical to the version involving real series.

Power series

Definition 2.29

A **complex power series** is a sum of the form $\sum\limits_{n=0}^{\infty} c_n z^n$, where $c_n \in \mathbb{C}$. The c_n are called the **coefficients** of the power series.

We treat such a power series as a function of z. Power series are central in the study of complex differentiable functions as we shall see in Chapter 13 onwards.

It is clear that, as in the case of real power series, a power series may not converge for all $z \in \mathbb{C}$. This motivates the following.

Definition 2.30

For a power series $\sum\limits_{n=0}^{\infty} c_n z^n$ the **radius of convergence**, R, is defined to be

$$R = \sup \{ |z| \ : \ \sum_{n=0}^{\infty} c_n z^n \text{ converges} \}.$$

Note that $R = \infty$ means that the series converges for all $z \in \mathbb{C}$.

Theorem 2.31

Suppose that the complex power series $\sum\limits_{n=0}^{\infty} c_n z^n$ has radius of convergence R. Then,

$$\sum_{n=0}^{\infty} c_n z^n \begin{cases} \text{converges absolutely for } |z| < R, \\ \text{diverges for } |z| > R. \end{cases}$$

The proof is similar to the real case.

Example 2.32

The power series $\sum_{n=0}^{\infty} z^n$ has radius of convergence 1.

To show this let us use the ratio test. Let $a_n = |z^n|$. We want $\sum_{n=0}^{\infty} a_n$ to converge (because if it does, then the series converges by Theorem 2.24: 'absolute convergence $\implies$ convergence'). So, we use the ratio test on this *real* series. We have

$$\left| \frac{a_{n+1}}{a_n} \right| = \frac{|z^{n+1}|}{|z^n|} = |z|.$$

As $n \to \infty$ we have $|z| \to |z|$, because there is no dependence on n. So by the ratio test the series $\sum a_n$ converges if $|z| < 1$, diverges if $|z| > 1$ and for $|z| = 1$ we don't know what will happen. So the radius of convergence is 1.

Example 2.33

Find the radius of convergence for $\sum_{n=1}^{\infty} \frac{(3z)^n}{n^2}$.

Solution: We use the ratio test on the series $\sum_{n=1}^{\infty} \left| \frac{(3z)^n}{n^2} \right|$. Let $a_n = \left| \frac{(3z)^n}{n^2} \right|$.

Then,

$$\begin{aligned}
\left| \frac{a_{n+1}}{a_n} \right| &= \left| \frac{(3z)^{n+1}}{(n+1)^2} \right| \Big/ \left| \frac{(3z)^n}{n^2} \right| \\
&= \left| \frac{(3z)^{n+1}}{(3z)^n} \Big/ \frac{n^2}{(n+1)^2} \right| \\
&= \left(\frac{n}{n+1} \right)^2 |3z| \\
&= 3 \left(\frac{n}{n+1} \right)^2 |z| \\
&\to 3.1.|z|, \text{ as } n \to \infty, \\
&= 3|z|.
\end{aligned}$$

By the ratio test the series converges absolutely, and hence converges, if $3|z| < 1$, i.e., $|z| < 1/3$, and diverges if $3|z| > 1$, i.e., $|z| > 1/3$.

Therefore, the radius of convergence is $1/3$.

Example 2.34

Find the radius of convergence of $\sum_{n=0}^{\infty} \left(\frac{1}{2} \right)^{n^2} z^n$.

Now $a_n = \left| \left(\frac{1}{2} \right)^{n^2} z^n \right|$. Then,

$$\left| \frac{a_{n+1}}{a_n} \right| = \left| \left(\frac{1}{2} \right)^{(n+1)^2} z^{n+1} \Big/ \left(\frac{1}{2} \right)^{n^2} z^n \right| = \left| \left(\frac{1}{2} \right)^{2n+1} z \right| = \left(\frac{1}{2} \right)^{2n+1} |z|$$

As $n \to \infty$, then $(1/2)^{2n+1} \to 0$. Thus, as z is fixed, $\lim_{n \to \infty} |a_{n+1}/a_n| = 0$. So, by the ratio test, the series converges absolutely for all z. That is, $R = \infty$.

Example 2.35
Find the radius of convergence of $\sum_{n=0}^{\infty} z^{n^2}$.

Let $a_n = \left| z^{n^2} \right|$. Then $|a_{n+1}/a_n| = |z|^{2n+1}$, which tends to zero if $|z| < 1$ and tends to infinity for $|z| > 1$. Thus the series converges for $|z| < 1$, diverges for $|z| > 1$, and so $R = 1$.

Common Error 2.36
A common mistake in applying the ratio test to a power series is to make $|z|$ go to infinity rather than n.

How to think like a mathematician 2.37
Given a power series, immediately ask 'What is its radius of convergence?'

Exercise 2.38
Show that $\sum_1^{\infty} z^n/n$ has radius of convergence 1.

It is known from the study of real series that the series in the last exercise converges for $z = -1$ and diverges for $z = 1$. Thus for a power series with radius of convergence R for $|z| = R$ we can have some values of z for which the series converges and some for which the series diverges.

Sin, cos, and exp are defined for all complex numbers

We can now show that the sine, cosine and exponential functions are defined on the whole of $\mathbb{C}$.

Example 2.39
The series

$$e^z = \sum_{n=0}^{\infty} \frac{z^n}{n!}$$

converges (absolutely) for all $z \in \mathbb{C}$:

For any $z \in \mathbb{C}$ let $a_n = \left| \dfrac{z^n}{n!} \right|$. We have

$$
\begin{aligned}
\left| \frac{a_{n+1}}{a_n} \right| &= \left| \frac{z^{n+1}}{z^n} \frac{n!}{(n+1)!} \right| \\
&= \frac{|z|}{n+1} \\
&\to 0 \text{ as } n \to \infty.
\end{aligned}
$$

So, by the ratio test, $\displaystyle\sum_{n=0}^{\infty} a_n = \sum_{n=0}^{\infty} \left| \frac{z^n}{n!} \right|$ converges for all $z \in \mathbb{C}$. Thus, by Theorem 2.24, the series $\displaystyle\sum_{n=0}^{\infty} \frac{z^n}{n!}$ converges for all $z \in \mathbb{C}$.

Hence, our definition e^z is good as it works for all $z \in \mathbb{C}$. That is, whatever z we use, we get a complex number from the series.

We can do the same for our complex sine function.

Example 2.40
The series $\sin z = \displaystyle\sum_{n=0}^{\infty} (-1)^n \frac{z^{2n+1}}{(2n+1)!}$ converges (absolutely) for all $z \in \mathbb{C}$.

Let $a_n = \left| (-1)^n \dfrac{z^{2n+1}}{(2n+1)!} \right|$. Then

$$
\begin{aligned}
\left| \frac{a_{n+1}}{a_n} \right| &= \left| \frac{z^{2n+3}}{(2n+3)!} \bigg/ \frac{z^{2n+1}}{(2n+1)!} \right| \\
&= \left| \frac{z^2}{(2n+3)(2n+2)} \right| \\
&= \frac{|z|^2}{(2n+3)(2n+2)} \\
&\to 0 \text{ as } n \to \infty.
\end{aligned}
$$

So by the ratio test the complex series converges absolutely, and hence converges.

Exercise 2.41
Prove that $\cos z$ converges for all $z \in \mathbb{C}$.

Exercises

(i) Determine whether the following are open sets in $\mathbb{C}$ or not.

(a) The interval $[0, 1]$,

(b) the interval $(0, 1)$,

(c) $\{z \in \mathbb{C} : |z - 2| < 5\}$,

(d) $\{z \in \mathbb{C} : |z - 2| > 5\}$,

(e) $\{z \in \mathbb{C} : |z - 2| < 5\} \cup \{z \in \mathbb{C} : |z - 2| > 5\}$,

(f) $\{z \in \mathbb{C} : |z + 1| \leq 4\}$,

(g) $\{z \in \mathbb{C} : |z + 1| = 4\}$,

(h) $z \in \mathbb{C}$ such that $\mathrm{Re}(z) > 5$,

(i) $z \in \mathbb{C}$ such that $\mathrm{Re}(z) \geq 5$,

(j) $z \in \mathbb{C}$ such that $\mathrm{Re}(z) > 5$ and $\mathrm{Im}(z)$ is negative,

(k) $z \in \mathbb{C}$ such that $\mathrm{Re}(z) > 5$ and $\mathrm{Im}(z)$ is non-positive,

(l) $\{z \in \mathbb{C} : |z + 1| < 6\} \cap \{z \in \mathbb{C} : \mathrm{Im}(z) > 2\}$.

Of those that are open which are domains?

(ii) Let D be the complement in $\mathbb{C}$ of a finite set of lines. Show that D is an open set.

(iii) Show that the limit of a complex sequence is unique.

(iv) Algebra of limits: Suppose that a_n and b_n are complex sequences. Show that if $\lim_{n \to \infty} a_n$ and $\lim_{n \to \infty} b_n$ exist, then

(a)
$$\lim_{n \to \infty} (a_n + b_n) = \lim_{n \to \infty} a_n + \lim_{n \to \infty} b_n,$$

(b)
$$\lim_{n \to \infty} (a_n b_n) = \lim_{n \to \infty} a_n \lim_{n \to \infty} b_n,$$

(c)
$$\lim_{n \to \infty} \left(\frac{a_n}{b_n} \right) = \frac{\lim_{n \to \infty} a_n}{\lim_{n \to \infty} b_n},$$

provided $b_n \neq 0$ for all n and $\lim_{n \to \infty} b_n \neq 0$.

(v) Prove that the Divergence test holds for complex series. That is, if $z_n \not\to 0$, then $\sum_{n=0}^{\infty} z_n$ diverges.

(vi) Find the radius of convergence for the following

$\quad$ (a) $\sum_{n=0}^{\infty}(-2)^n z^n$,

$\quad$ (b) $\sum_{n=0}^{\infty}\left(\frac{1}{2}\right)^{n^2} z^n$,

$\quad$ (c) $\sum_{n=0}^{\infty} z^{n^2}$,

$\quad$ (d) $\sum_{n=0}^{\infty} n! z^n$,

$\quad$ (e) $\sum_{n=1}^{\infty}(\log n)^2 z^n$.

(vii) Suppose that $\sum_{n=0}^{\infty} a_n$ converges. Show that

$$\overline{\sum_{n=0}^{\infty} a_n} = \sum_{n=0}^{\infty} \overline{a_n}.$$

(viii) Suppose that $a(z) = \sum_{n=0}^{\infty} a_n z^n$ has radius of convergence R_1 and $b(z) = \sum_{n=0}^{\infty} b_n z^n$ has radius of convergence R_2. Prove that for $c_n = \sum_{k=0}^{n} a_k b_{n-k}$

$$\sum_{n=0}^{\infty} c_n z^n = a(z)b(z)$$

and this has radius of convergence at least $\min\{R_1, R_2\}$.

(ix) In Exercise 2.38 we see that different behaviour can occur on the radius of convergence. The following gives a condition for convergence. Suppose that a_n is real, $a_n \geq 0$ and that that the series $\sum_{n=0}^{\infty} a_n R^n$ converges for some $R > 0$. Prove that the series $\sum_{n=0}^{\infty} a_n z^n$ converges for all $|z| = R$.

(x) Show that

$\quad$ (a) $\displaystyle\sum_{n=0}^{\infty} z^n = \frac{1}{1-z}$ for $|z| < 1$,

$\quad$ (b) $\displaystyle\sum_{n=1}^{\infty} n z^{n-1} = \frac{1}{(1-z)^2}$ for $|z| < 1$.

Generalize the result in (b). Differentiation is not allowed as it has not been defined yet.

Summary

- A sequence of complex numbers z_n converges to $z \in \mathbb{C}$, if for all $\varepsilon > 0$ there exists N such that $|z_n - z| < \varepsilon$ for all $n \geq N$.

- Paradigm: Complex analysis is developed by reducing to real analysis, often through taking the modulus and argument or taking real and imaginary parts.

- We define exponential, sine and cosine by power series.

- A complex series $\sum_{k=0}^{\infty} z_k$ converges if the sequence s_n formed by its partial sums $s_n = \sum_{k=0}^{n} z_k$ converges.

- $\sum_{k=0}^{\infty} z_k = \sum_{k=0}^{\infty} x_k + i \sum_{k=0}^{\infty} y_k$.

- Absolute convergence implies convergence: If $\sum_{k=0}^{\infty} |z_k|$ converges, then $\sum_{k=0}^{\infty} z_k$ converges.

- Suppose that $\sum_{k=0}^{\infty} z_k$ converges absolutely. Then $\left|\sum_{k=0}^{\infty} z_k\right| \leq \sum_{k=0}^{\infty} |z_k|$.

- Suppose that the series $\displaystyle\sum_{n=0}^{\infty} a_n$ and $\displaystyle\sum_{n=0}^{\infty} b_n$ are convergent such that at least one is absolutely convergent. Then $\sum_{n=0}^{\infty} c_n = \left(\sum_{n=0}^{\infty} a_n\right)\left(\sum_{n=0}^{\infty} b_n\right)$ where
$$c_n = \sum_{k=0}^{n} a_k b_{n-k}.$$

- Radius of convergence: $R = \sup\{|z| \; : \; \sum_{n=0}^{\infty} c_n z^n \text{ converges}\}$. Also,
$$\sum_{n=0}^{\infty} c_n z^n \begin{cases} \text{converges absolutely for } |z| < R, \\ \text{diverges for } |z| > R, \\ \text{may or may not converge for } z \text{ such that } |z| = R. \end{cases}$$

- Apply the ratio test, comparison test, etc, to the modulus of terms of a complex series to determine convergence.

CHAPTER 3

The Exponential and Complex Logarithm

Having seen that the definition of the complex exponential function is a good one, we shall now show that it has some but not all of the properties of the real exponential function. For example, like the real version, the exponential is never zero, but unlike the real one, it is not injective. Lack of injectivity implies that it does not have an inverse and, consequently, we have to define the complex logarithm function with care.

Properties of the exponential

Theorem 3.1
The complex exponential function has the following properties:

   (i) $e^{\bar{z}} = \overline{e^z}$, *for all* $z \in \mathbb{C}$.

  (ii) $e^{iz} = \cos z + i \sin z$, *for all* $z \in \mathbb{C}$.

 (iii) $e^{z+w} = e^z e^w$, *for all* $z, w \in \mathbb{C}$.

 (iv) $e^z \neq 0$, *for all* $z \in \mathbb{C}$.

  (v) $e^{-z} = 1/e^z$, *for all* $z \in \mathbb{C}$.

 (vi) $e^{nz} = (e^z)^n$, *for all* $z \in \mathbb{C}$ *and* $n \in \mathbb{Z}$.

(vii) $|e^z| = e^{Re(z)}$, *for all $z \in \mathbb{C}$.*

(viii) $|e^z| \leq e^{|z|}$, *for all $z \in \mathbb{C}$.*

(ix) $|e^{iy}| = 1$, *for all $y \in \mathbb{R}$.*

Proof. (i) We have

$$e^{\bar{z}} = \sum_{n=0}^{\infty} \frac{(\bar{z})^n}{n!} = \sum_{n=0}^{\infty} \frac{\overline{(z^n)}}{n!} = \overline{\sum_{n=0}^{\infty} \frac{z^n}{n!}} = \overline{e^z}.$$

(ii) Exercise.

(iii) Since the series for e^z is absolutely convergent we can use Theorem 2.28:

$$
\begin{aligned}
e^z e^w &= \left(\sum_{n=0}^{\infty} \frac{z^n}{n!} \right) \left(\sum_{n=0}^{\infty} \frac{w^n}{n!} \right) \\
&= \sum_{n=0}^{\infty} \sum_{k=0}^{n} \frac{z^k}{k!} \frac{w^{n-k}}{(n-k)!}, \text{ by Theorem 2.28,} \\
&= \sum_{n=0}^{\infty} \frac{1}{n!} \sum_{k=0}^{n} \frac{n!}{k!(n-k)!} z^k w^{n-k} \\
&= \sum_{n=0}^{\infty} \frac{1}{n!} (z+w)^n, \text{ by the binomial theorem,} \\
&= e^{z+w}.
\end{aligned}
$$

(iv) Note that e^z and e^{-z} both exist. We have

$$
\begin{aligned}
e^z e^{-z} &= e^{z-z} \text{ by (iii),} \\
&= e^0 \\
&= 1, \text{ by calculation.}
\end{aligned}
$$

Thus, e^z cannot be zero.

(v) This is obvious from the proof of (iv).

(vi) Follows from repeated application of (iii).

(vii) We have

$$
\begin{aligned}
|e^z|^2 &= e^z \overline{e^z} \text{ by definition,} \\
&= e^z e^{\bar{z}} \text{ by (i),} \\
&= e^{z+\bar{z}} \text{ by (iii),} \\
&= e^{2Re(z)} \\
&= \left(e^{Re(z)} \right)^2 \text{ by (vi).}
\end{aligned}
$$

As both $|e^z|$ and $e^{\mathrm{Re}(z)}$ are real and positive we deduce that (vii) is true.

(viii) Follows from (vii) as $\mathrm{Re}(z) \le |z|$ and as the real exponential is an increasing function.

(ix) From (vii) we get $|e^{iy}| = e^{\mathrm{Re}(iy)} = e^0 = 1$. □

From these we can deduce some standard results.

Corollary 3.2

(i) Euler's identity: $e^{\pi i} = -1$.

(ii) De Moivre's Theorem: For all $\theta \in \mathbb{R}$, $(\cos\theta + i\sin\theta)^n = \cos n\theta + i\sin n\theta$.

The proofs are left as simple exercises. Part (i) is one of the **best theorems in mathematics**. If we square both sides we get $e^{2\pi i} = 1$. This form is interesting as it relates so many different important numbers, e, $\sqrt{-1}$, π, and of course 1 and 2, in a simple expression.

Warning! 3.3

We have **not** shown that $e^{zw} = (e^z)^w$ for all $z, w \in \mathbb{C}$. This is because we have not yet *defined* a^b for all complex a and b. Consider $z = 2\pi i$ and $w = i$. Then $(e^z)^w = (e^{2\pi i})^i = 1^i$. What could 1^i be?

Exercise 3.4

Prove that

$$\sin z = \frac{e^{iz} - e^{-iz}}{2i} \qquad \text{and} \qquad \cos z = \frac{e^{iz} + e^{-iz}}{2}.$$

We can now derive properties of sine and cosine from properties of the exponential (and vice versa).

Another property of the complex exponential is that it is periodic.

Theorem 3.5

For any complex numbers z and w we have

$$e^z = e^w \iff z - w = 2\pi i n \text{ for some } n \in \mathbb{Z}.$$

Proof. [$\Rightarrow$] Let $z - w = x + iy$ where x and y are real. Then,

$$\begin{aligned}
e^z = e^w &\iff e^{z-w} = 1 \\
&\iff e^{x+iy} = 1 &(*)\\
&\implies |e^{x+iy}| = 1, \text{ (the implication does not reverse!)} \\
&\iff |e^x||e^{iy}| = 1 \\
&\iff e^x = 1 \\
&\iff x = 0.
\end{aligned}$$

By (∗) we know that $e^{x+iy} = 1$, so $e^{iy} = 1$ as $x = 0$. Then,

$$
\begin{aligned}
e^{iy} = 1 \quad &\Longleftrightarrow \quad \cos y + i \sin y = 1 \\
&\Longleftrightarrow \quad \cos y = 1 \text{ and } \sin y = 0 \\
&\Longleftrightarrow \quad y = 2\pi n \text{ for some } n \in \mathbb{Z}.
\end{aligned}
$$

So $z - w = x + iy = 0 + i.2\pi n = 2\pi in$.

[⇐] Suppose that $z - w = 2\pi in$ for some $n \in \mathbb{Z}$. Then, $e^{z-w} = e^{2\pi in} = \left(e^{2\pi i}\right)^n = 1^n = 1$. But from the working in the earlier part of the proof we know this is equivalent to $e^z = e^w$. □

Example 3.6

Probably, the simplest example is that $e^0 = 1$ and $e^{2\pi i} = 1$.

An important consequence of the theorem is that in contrast to the real case the complex exponential function is not injective. This has serious implications for defining the inverse of e^z, i.e., defining a logarithmic function.

Definition of the complex logarithm

Before we attempt to define an inverse for the exponential let us investigate solutions of $e^w = z$ as this must be the basis of any definition.

Proposition 3.7

For complex numbers w and $z \neq 0$ we have

$$
e^w = z \iff w = \log_e |z| + i(\operatorname{Arg}(z) + 2k\pi), \text{ for } k \in \mathbb{Z}.
$$

Proof. Let $w = x + iy$. Then

$$
\begin{aligned}
e^w = z \quad &\Longleftrightarrow \quad e^x e^{iy} = |z| e^{i \operatorname{Arg}(z)} \\
&\Longleftrightarrow \quad |e^x| = |z| \text{ and } y = \operatorname{Arg}(z) + 2k\pi \\
&\Longleftrightarrow \quad x = \log_e |z| \text{ and } y = \operatorname{Arg}(z) + 2k\pi.
\end{aligned}
$$

□

Example 3.8

Solve $e^w = 1 + i\sqrt{3}$.

Solution: Let $z = 1 + i\sqrt{3}$. The modulus of z is $|z| = \sqrt{1^2 + \sqrt{3}^2} = \sqrt{4} = 2$. By drawing a picture (or through careful use of a calculator) we can see that $\operatorname{Arg}(z) = \dfrac{\pi}{3}$. So

$$
w = \log_e 2 + i\left(\frac{\pi}{3} + 2n\pi\right), \ n \in \mathbb{Z}
$$

gives the solutions for the equation.

Common Error 3.9

Don't forget the $2k\pi$ with the argument.

Now let us attempt to define an inverse for the exponential function. Let $D = \{z \in \mathbb{C} \mid -\pi < \text{Im}(z) < \pi\}$. Then D is an open set and the exponential function is injective on D by Theorem 3.5. Hence, the restriction of the exponential to D will be a bijection onto its image.

Lemma 3.10

The image of D under the exponential function is $\mathbb{C}$ without the non-positive real axis. That is,

$$\{e^z \mid z \in D\} = \mathbb{C}\backslash\{x \in \mathbb{R} \mid x \leq 0\}.$$

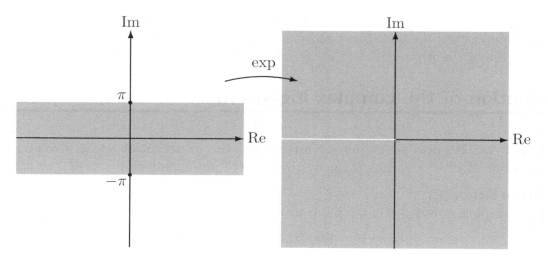

Proof. Suppose that $z \in \mathbb{C}\backslash\{x \in \mathbb{R} \mid x \leq 0\}$. Then, $z = |z|e^{i\,\text{Arg}(z)}$ where $-\pi < \text{Arg}(z) < \pi$. Let $w = \log_e |z| + i\,\text{Arg}(z)$. Then $w \in D$ and $e^w = z$.

If z is in the non-positive real axis, then $\text{Arg}(z) = \pi$ and hence any w such that $e^w = z$ is such that $w \notin D$. $\qquad\square$

This lemma tells us that $\exp : D \to \mathbb{C}\backslash\{x \in \mathbb{R} \mid x \leq 0\}$ is a bijection. (Here $\exp$ is just the function given by $\exp(z) = e^z$.) Hence, we make the following definition.

Definition 3.11

Let $z \neq 0$ be a complex number such that $\text{Arg}(z) \neq \pi$. The **complex logarithm**, denoted Log, is defined by

$$\text{Log}\, z = \log |z| + i\,\text{Arg}(z).$$

The logarithm has the following properties.

Theorem 3.12

(i) For z a positive real number, then $\operatorname{Log} z = \log_e z$.

(ii) For all z with $-\pi < \operatorname{Im}(z) < \pi$, then $\operatorname{Log} e^z = z$.

(iii) For all z not on the non-positive real axis, $e^{\operatorname{Log} z} = z$.

(iv) If all z_1, z_2 and $z_1 z_2$ do not have argument π, then we have

$$\operatorname{Log}(z_1 z_2) = \operatorname{Log} z_1 + \operatorname{Log} z_2 + 2n\pi i \text{ where } n = -1, 0, \text{ or } 1.$$

The proofs are left as exercises.

How to think like a mathematician 3.13

Obviously, in analogy with the real case, we would like $\operatorname{Log}(z_1 z_2) = \operatorname{Log} z_1 + \operatorname{Log} z_2$ to be true. Note that the theorem does not actually rule out the possibility that n is always zero in part (iv) of the theorem. However, examples do rule this out. This is also left as an exercise.

We can use the logarithm to solve equations involving sines and cosines.

Example 3.14

Solve the equation $\cos z = 4$.

Solution: We can rewrite this as $\dfrac{e^{iz} + e^{-iz}}{2} = 4$. Let $w = e^{iz}$. Then the equation becomes $\dfrac{1}{2}\left(w + \dfrac{1}{w}\right) = 4$. So,

$$
\begin{aligned}
w^2 + 1 &= 8w \\
w^2 - 8w + 1 &= 0 \\
w &= \frac{8 \pm \sqrt{64 - 4}}{2} \\
&= \frac{8 \pm \sqrt{60}}{2} \\
&= 4 \pm \sqrt{15}.
\end{aligned}
$$

Now, $e^{iz} = w = 4 \pm \sqrt{15}$, so

$$\begin{aligned}
iz &= \ln|w| + i(\mathrm{Arg}(w) + 2n\pi), \ n \in \mathbb{Z}, \\
&= \ln\left|4 \pm \sqrt{15}\right| + i\,(0 + 2n\pi) \\
z &= \frac{1}{i}\left(\ln\left|4 \pm \sqrt{15}\right| + 2n\pi i\right) \\
&= -i\ln\left|4 \pm \sqrt{15}\right| + (-i)2n\pi i \\
&= -2n\pi - i\ln\left(4 \pm \sqrt{15}\right).
\end{aligned}$$

(The last equality is true because $\ln(4 + \sqrt{15}) > 0$ and $\ln(4 - \sqrt{15}) > 0$.)

How to think like a mathematician 3.15
Note that in the above example we replaced e^{iz} with another complex number w, because we could then get a polynomial equation.

Exercises

Exercises 3.16

(i) Prove Theorem 3.1(ii).

(ii) Draw on an Argand diagram the image of the map $\gamma(t) = re^{it}$ where $t \in \mathbb{R}$ and $r > 0$.

(iii) Let exp be the exponential function given by $\exp(z) = e^z$. Let X be a horizontal line in the complex plane. Draw $\exp(X)$. Let X now be a vertical line. Draw $\exp(X)$ for this.

(iv) Show that there exists $z \in \mathbb{C}\backslash\mathbb{R}$ such that $e^z \in \mathbb{R}$ and $e^z < 0$. Find all such z and draw on an Argand diagram.

(v) Show that

$$\begin{aligned}
\sin z = 0 &\iff z = k\pi, \ k \in \mathbb{Z}, \\
\cos z = 0 &\iff z = \frac{1}{2}(2k+1)\pi, \ k \in \mathbb{Z}.
\end{aligned}$$

These mean that the zeroes of sine and cosine lie on the real line.

(vi) We can define the **tangent function**, tan, to be

$$\tan(z) = \frac{\sin(z)}{\cos(z)}.$$

Find the largest open set upon which tan is defined.

(vii) Find all the solutions of $e^z = 1$ and $e^z = -1$.

(viii) Prove that $\dfrac{\sin z}{z} \to 1$ as $z \to 0$.

(ix) Express $\cos^2(z)$ as a power series in z, giving all terms up to z^6.

(x) Show that

(a) $\sin(z \pm w) = \sin z \cos w \pm \cos z \sin w$, for all $z, w \in \mathbb{C}$;

(b) $\cos(z \pm w) = \cos z \cos w \mp \sin z \sin w$, for all $z, w \in \mathbb{C}$.

Hence, deduce that $\sin^2 z + \cos^2 z = 1$ for all $z \in \mathbb{C}$.

(xi) Show that

$$\sin(z + 2\pi) = \sin(z), \qquad \cos(z + 2\pi) = \cos(z),$$

$$\sin\left(\frac{\pi}{2} - z\right) = \cos z, \qquad \cos\left(\frac{\pi}{2} - z\right) = \sin z.$$

(xii) Show that $\sin \overline{z} = \overline{\sin z}$ and $\cos \overline{z} = \overline{\cos z}$. Is $\tan \overline{z} = \overline{\tan z}$? If so, then prove it. If not, then give a counterexample.

(xiii) We define the **complex hyperbolic sine and cosine** by

$$\sinh z = \frac{1}{2}(e^z - e^{-z}) \text{ and } \cosh z = \frac{1}{2}(e^z + e^{-z}).$$

(a) Prove that $\sin iz = i \sinh z$ and $\cos iz = \cosh z$ for all $z \in \mathbb{C}$.

(b) Prove that $|\cos(x + iy)|^2 = \cos^2 x + \cosh^2 y$ for all $x, y \in \mathbb{R}$. Hence, deduce that $|\cos(z)|$ is an unbounded function on $\mathbb{C}$.

(c) Write sinh and cosh as power series.

(d) Show that $|\sinh(z)| \le e^{|z|}$ and $|\cosh(z)| \le e^{|z|}$.

(xiv) Simplify

(a) $\text{Log}(i)$, (b) $\text{Log}(i + i)$, (c) $\text{Log}(-3i)$,

(xv) Prove the statements in Theorem 3.12. Give an example to show that n is not always zero in part (iv).

(xvi) Prove that $\text{Log}(1/z) = -\text{Log}(z)$ for all z with $\text{Arg}(z) \neq \pi$.

(xvii) Solve the following equations.

(a) $z^3 = 1 + 3i$.

(b) $e^z = \sqrt{3} + i$

(c) $e^{2iz} = i$. (It's not $i(\pi/2 + 2k\pi)$.)

(d) $\cos z = 3$.

(e) $\sin z = 2$.

(xviii) Which properties of the real exponential do not hold for the complex exponential?

Summary

❏ The complex exponential has some but not all of the properties of the real exponential.

❏ For complex numbers w and $z \neq 0$ we have

$$e^w = z \iff w = \log_e |z| + i(\text{Arg}(z) + 2k\pi), \text{ for } k \in \mathbb{Z}.$$

❏ Let $z \neq 0$ be a complex number such that $\text{Arg}(z) \neq \pi$. The complex logarithm, denoted Log, is defined by

$$\text{Log } z = \log |z| + i\,\text{Arg}(z).$$

❏ The sine and cosine addition formula hold. (See exercises.)

CHAPTER 4

Calculus of Real Variable Complex-Valued Functions

Complex analysis is the generalization of calculus from real functions to complex functions. To achieve this we first study a stepping stone between the two: we generalise integration and differentiation to *complex*-valued functions of a *real* variable. Fortunately, this behaves very much like real-valued functions of a real variable.

Definition 4.1
A **complex-valued function of a real variable** is a map $f : S \to \mathbb{C}$, where $S \subseteq \mathbb{R}$.

Example 4.2
If $f(t) = (2 + 3i)t^3$, $t \in \mathbb{R}$, then $f(1) = 2 + 3i \in \mathbb{C}$.

Such a function is different to a complex function. A complex-valued function of a real variable takes a real number and produces a complex number. A complex function takes a complex number from an open set in $\mathbb{C}$ and produces a complex number.

Differentiating complex-valued real variable functions

Let (a, b) be an interval in $\mathbb{R}$ and let $f : (a, b) \to \mathbb{C}$ be a complex-valued functions of a real variable. Then the derivative of f is defined in the same way as for real calculus:

$$f'(t) = \lim_{h \to 0} \frac{f(t+h) - f(t)}{h}, \quad (h \in \mathbb{R}),$$

For example, suppose that $f : \mathbb{R} \to \mathbb{C}$ is given by $f(t) = (1 + 3i)t^2$, then $f'(t) = 2(1 + 3i)t$.

Any complex-valued function of a real variable, $f : S \to \mathbb{C}$ can be written as $f(t) = g(t) + ih(t)$ where g and h are real functions. In this case we have $f'(t) = g'(t) + ih'(t)$. From this we can show that the derivative of a sum is the sum of derivatives. That is, for $f = f_1 + f_2$, we have $f'(t) = f_1'(t) + f_2'(t)$. Similarly the product and quotient rules hold as well as the rules of differentiation of elementary functions.

Example 4.3

For all $c \in \mathbb{C}$,

$$\frac{d}{dt} ct^n = nct^{n-1},$$

$$\frac{d}{dt} e^{ct} = ce^{ct},$$

$$\frac{d}{dt} \sin(ct) = c\cos(ct),$$

$$\frac{d}{dt} \cos(ct) = -c\sin(ct).$$

Examples 4.4

(i) Let $g(s) = (5 + i)\sin(2s) + 3\cos t - e^{is}$. Then

$$g'(s) = 2(5 + i)\cos(2s) - ie^{is}.$$

(ii) Let $\phi(x) = 3x^3 + 2ix - i + \tan((4 + 2i)x)$. Then,

$$\phi'(x) = 6x + 2i + (4 + 2i)\left(1 + \tan\left((4 + 2i)x\right)^2\right).$$

(Recall that we can define a complex version of tan by $\tan(z) = \dfrac{\sin(z)}{\cos(z)}$.)

Remark 4.5

Differentiation of a complex function with respect to a real variable is not the same as differentiation with respect to a complex variable. We will define the latter in Chapter 8.

We can also define continuity in the standard way.

Definition 4.6

Suppose $f : S \to \mathbb{C}$ is a complex-valued function of a real variable. Then f is **continuous at** $x \in S$ if $\lim_{t \to x} f(t) = f(x)$. We say that f is **continuous** if f is continuous for all $x \in S$.

For certain situations it is useful to ensure that the derivative is well-behaved.

Definition 4.7

Suppose $f : (a, b) \to \mathbb{C}$ is a complex-valued function of a real variable. Then f is **continuously differentiable** if f is differentiable and f' is continuous.

Integrating complex-valued real variable functions

We've done differentiation, now we shall integrate complex-valued functions with respect to a real variable.

Definition 4.8

Suppose that $g : [a, b] \to \mathbb{C}$ is a complex-valued function of a real variable given by $g(t) = u(t) + iv(t)$. We say that g is **complex Riemann integrable** if both u and v are Riemann integrable as real functions. We define $\int_a^b g(t)\, dt$ to be

$$\int_a^b g(t)\, dt = \int_a^b u(t)\, dt + i \int_a^b v(t)\, dt.$$

Example 4.9

$$
\begin{aligned}
\int_a^b e^{4it}\, dt &= \int_a^b \cos 4t\, dt + i \int_a^b \sin 4t\, dt \\
&= \frac{1}{4}\left[\sin 4b - \sin 4a\right] + \frac{i}{4}\left[-\cos 4b + \cos 4a\right] \\
&= \frac{1}{4i}\left(i(\sin 4b - \sin 4a) - (-\cos 4b + \cos 4a)\right) \\
&= \frac{1}{4i}\left(e^{4ib} - e^{4ia}\right) \\
&= \frac{i}{4}\left(e^{4ia} - e^{4ib}\right).
\end{aligned}
$$

Many properties of the complex Riemann integral can be derived from the corresponding properties for the real version of Riemann integration by considering real and imaginary parts. For example, there is a version of the Fundamental Theorem of Calculus:

Proposition 4.10 (Fundamental Theorem of Calculus)
Let $g : [a, b] \to \mathbb{C}$ be a continuous function and $G : (a, b) \to \mathbb{C}$ is such that $G' = g$. Then,

$$\int_a^b g(t)\, dt = G(b) - G(a).$$

Proof. Let $G(t) = u(t) + iv(t)$ where u and v are real functions. As g is continuous u' and v' are continuous on (a, b) and hence are Riemann integrable. Therefore,

$$\begin{aligned}
\int_a^b g(t)\, dt &= \int_a^b G'(t)\, dt \\
&= \int_a^b u'(t)\, dt + i \int_a^b v'(t)\, dt \\
&= [u(b) - u(a)] + i\,[v(b) - v(a)] \\
&= G(b) - G(a).
\end{aligned}$$

$\square$

Other standard methods, such as substitution also work.

Obviously, separating functions into real and imaginary parts is tedious. Fortunately, as for differentiation, we can use standard integrals, replacing real constants by complex ones. It is not difficult to prove the following.

Examples 4.11
For $c \in \mathbb{C}$ and $a, b \in \mathbb{R}$ we have

$$\begin{aligned}
\int_a^b ct^n\, dt &= \left[c\frac{t^{n+1}}{n+1} \right]_a^b, \\
\int_a^b e^{ct}\, dt &= \left[\frac{e^{ct}}{c} \right]_a^b, \\
\int_a^b \sin ct\, dt &= \left[-\frac{1}{c}\cos(ct) \right]_a^b, \\
\int_a^b \cos ct\, dt &= \left[\frac{1}{c}\sin(ct) \right]_a^b.
\end{aligned}$$

Exercises

Exercises 4.12

  (i) Prove the product and quotient rules for differentiation.

  (ii) Prove the equalities in Example 4.3. (Hint: The derivatives of sine and cosine will follow from that of the exponential when we use Exercise 3.4 To find the derivative of the exponential we can write e^{ct} as $e^{(a+bi)t} = e^{at}e^{ibt}$, i.e., as a product for some real a and b. Then use $e^{ibt} = \cos bt + i \sin bt$ and the product rule.)

  (iii) Stating and proving a version of the chain rule is not so elementary. Explain why. What is the closest statement to the chain rule that you can state?

  (iv) Calculate the following integrals:

    (a) $\int_0^1 |t + it|(1 + i)\, dt$,

    (b) $\int_1^3 (t + it^2)^2\, dt$

    (c) $\int_0^2 t^2 - it^3 - \cos(2t)\, dt$,

    (d) $\int_0^{\pi/2} ie^{e^{it}} e^{it}\, dt$.

  (v) Prove the equalities in Examples 4.11.

  (vi) Use the Fundamental Theorem of Calculus (Proposition 4.10) to verify the calculation in Example 4.9. Which method do you find easiest?

Summary

❏ We can integrate and differentiate a complex-valued function of a real variable in the same way as a real-valued function of a real variable.

❏ A version of the Fundamental Theorem of Calculus holds for complex valued functions of a real variable.

CHAPTER 5

Contour Integration

In this chapter we meet the central object of complex analysis: **the contour integral**. This is a fairly abstract concept, the *meaning* of which usually takes time to understand. Fortunately, it is easy to *calculate* as it has similar properties to integration of one real variable.

It is important to realize that we will not just be doing abstraction for abstraction's sake. As way of motivation, consider the following real integral:

$$\int_0^{2\pi} e^{\cos\theta} \cos(n\theta - \sin\theta)\, d\theta.$$

This is a seriously nasty integral! Imagine trying to evaluate it via the standard methods of real analysis. We shall see in Chapter 20 that this is extremely easy to determine using contour integration.

Contours

First, we define contours in the complex plane.

Definition 5.1
Let $\gamma : [a, b] \to \mathbb{C}$ be a map and suppose that $a = a_0 < a_1 < a_2 < \cdots < a_n = b$. We say that γ is a **contour** (also called a **path**) if

  (i) γ is continuous,

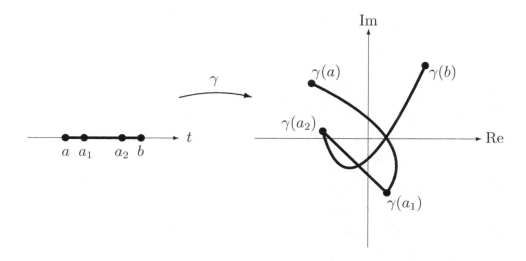

Figure 5.1: The image of the contour γ is on the right.

(ii) $\gamma|[a_{j-1}, a_j]$ is differentiable for $j = 1, \ldots, n$, that is, differentiable on the open interval (a_{j-1}, a_j) with left and right derivatives at end points,

(iii) $\gamma'|[a_{j-1}, a_j]$ is continuous on $[a_{j-1}, a_j]$ for $j = 1, \ldots, n$.

We denote the image of γ by γ^* and we say γ is **closed** if $\gamma(a) = \gamma(b)$.

The left and right derivatives of γ at a_j may differ. To avoid the resulting ambiguity we define $\gamma'(a_j)$ to be right derivative arising from $[a_j, a_{j+1}]$.

Definition 5.2
We can draw an arrow on a diagram to show which direction the curve goes in. This is called the **orientation of the contour**.

A contour γ is called **simple** if it does not cross itself, i.e., $\gamma(t_1) \neq \gamma(t_2)$ for all $t_1 \neq t_2$, except possibly for $\gamma(a) = \gamma(b)$. For a closed contour with no orientation specified, then by convention it is traversed anticlockwise.

Warning! 5.3
A contour is not a complex function. It is a complex-valued function of a real variable. Its image is usually some curve in the plane. (In theory it could be a simple point.)

Examples 5.4
(i) Straight line from α to β: This is $\gamma : [0, 1] \to \mathbb{C}$ given by $\gamma(t) = \alpha + t(\beta - \alpha)$.

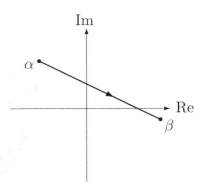

(ii) Circle of radius r based at w: This is $C_r : [0, 2\pi] \to \mathbb{C}$ given by $C_r(t) = w + re^{it}$.

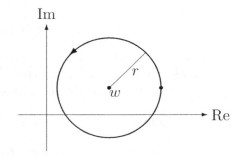

(iii) Circular arc of radius r based at w: $A_r(t) = w + re^{it}$, $\theta_1 \leq t \leq \theta_2$. (So $0 \leq t \leq 2\pi$ gives the circle above).

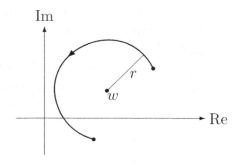

(iv) Let $\alpha : [-1, \pi/2] \to \mathbb{C}$ be given by

$$\alpha(t) = \begin{cases} t + 1, & \text{for } -1 \leq t \leq 0, \\ e^{it} & \text{for } 0 \leq t \leq \pi/2. \end{cases}$$

Draw the image:

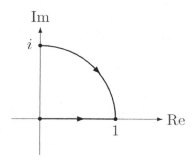

(v) Let $\gamma : [0, 4\pi] \to \mathbb{C}$ be given by $\gamma(t) = e^{it}$. Then the image of γ is the unit circle centred at zero. The contour goes round the circle twice.

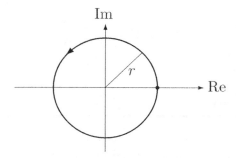

(vi) Let $\gamma : [0, 2] \to \mathbb{C}$ be defined by

$$\gamma(t) = \begin{cases} (1 + i)t^2, & \text{for } 0 \leq t \leq 1, \\ (2 - t) + (3t^2 - 10t + 8)i, & \text{for } 1 \leq t \leq 2. \end{cases}$$

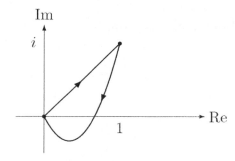

Note that for $\gamma(t) = (1+i)t^2$ part that the expression is quadratic but the object that appears in the picture is linear, i.e., a line!

How to think like a mathematician 5.5
Given a contour, draw its image.

Common Error 5.6
There is often confusion between a contour and its image. This is not surprising given the traditional notation used in complex analysis where $z \in \gamma$ is used for $z \in \gamma([a,b])$, i.e., z in the image of γ. We use the notation γ^* for the image of γ.

The important point to remember is that a contour is not a set of points in the complex plane, it is a map. For example, consider the contours (i) and (v) above, taking $w = 0$ and $r = 1$ in (i). They have the same image, the unit circle. However, the contours are different, one maps from $[0, 2\pi]$, the other $[0, 4\pi]$. The contours are different. The images are the same.

Remark 5.7
Since contours are complex-valued functions of a real variables, we can differentiate and integrate them.

Contour Integration

We define continuity for complex functions in an unsurprising way.

Definition 5.8
Suppose $f : D \to \mathbb{C}$ is a complex function on the open set D. Then f is **continuous at** $w \in D$ if $\lim_{z \to w} f(z) = f(w)$. We say f is **continuous** if f is continuous at all $c \in D$.

Note that this is equivalent to continuity of the real valued functions of two variables given by $\mathrm{Re}(f)$ and $\mathrm{Im}(f)$.

Now for the central definition in complex analysis.

Definition 5.9
Let D be an open set in $\mathbb{C}$. Let $f : D \to \mathbb{C}$ be a continuous complex function and $\gamma : [a,b] \to D$ be a contour. Then, the **contour integral of f along γ**, denoted $\int_\gamma f(z)\,dz$ is defined by

$$\int_\gamma f(z)\,dz = \int_a^b f(\gamma(t))\gamma'(t)\,dt.$$

Example 5.10

Let $\gamma(t) = t + it^2$ for $0 \leq t \leq 2$ and $f(z) = z$. Then, $\gamma'(t) = 1 + 2it$, and $f(\gamma(t)) = t + it^2$.

$$
\begin{aligned}
\int_\gamma f &= \int_0^2 (t + it^2)(1 + 2it)\, dt \\
&= \int_0^2 t + 2it^2 + it^2 + 2i^2 t^3 \, dt \\
&= \int_0^2 t + 3it^2 - 2t^3 \, dt \\
&= \left[\frac{1}{2}t^2 + \frac{3it^3}{3} - \frac{2t^4}{4} \right]_0^2 \\
&= \left[\frac{1}{2}t^2 + it^3 - \frac{t^4}{2} \right]_0^2 \\
&= \frac{1}{2}2^2 + i2^3 - \frac{2^4}{2} \\
&= -6 + 8i.
\end{aligned}
$$

Example 5.11

Let $\gamma(t) = 2 + it^2$ for $0 \leq t \leq 1$ and $f(z) = z^2$. Then,

$$
\begin{aligned}
\int_\gamma z^2 \, dz &= \int_0^1 (2 + it^2)^2 (2it)\, dt \\
&= \int_0^1 \frac{d}{dt}\left(\frac{(2 + it^2)^3}{3} \right) dt \\
&= \left[\frac{(2 + it^2)^3}{3} \right]_0^1 \\
&= \frac{(2 + i.1^2)^3}{3} - \frac{(2 + i.0^2)^3}{3} \\
&= \frac{1}{3}\left((2 + i)^3 - 8 \right) \\
&= -2 + \frac{11}{3}i.
\end{aligned}
$$

I strongly suggest that you attempt the following exercises now. A firm grasp of contour integrals is essential for the rest of this book. They are the central object in complex analysis.

Exercise 5.12

Draw the contours and calculate the integrals of the functions along the contours.

(i) $f_1(z) = \text{Re}(z)$ and $\gamma_1(t) = t$, $0 \leq t \leq 1$.

(ii) $f_2(z) = \text{Re}(z)$ and $\gamma_2(t) = t + it$, $0 \leq t \leq 1$.

(iii) $f_3(z) = \text{Re}(z)$ and $\gamma_3(t) = 1 - t + i(1 - t)$, $0 \leq t \leq 1$.

(iv) $f_4(z) = 1/z$ and $\gamma_4(t) = 2e^{-it}$, $0 \leq t \leq \pi$.

(v) $f_5(z) = z^2$ and $\gamma_5(t) = e^{it}$, $0 \leq t \leq \pi/2$.

Can you explain the connection between the results in (ii) and (iii)? Can you make any conjectures, say, involving $f(z) = z^n$ in (v)?

Remarks 5.13

(i) A contour integral is a complex number.

(ii) It is not obvious a priori that the integral in the definition of contour integral always exists. Let's remedy this. First, $f(\gamma(t))$ and $\gamma'(t)$ are complex-valued functions of a real variable, and hence so is their product. We assume that f is continuous and from the definition of a contour we know that γ' is continuous except possibly at a finite number of points. Hence the product is Riemann integrable. That is, the integral always exists.

When γ' has discontinuities, then we divide the integral into pieces as in the following.

Example 5.14
Let γ be as in Example 5.4(v). Find $\int_\gamma z^5 \, dz$.

$$
\begin{aligned}
\int_\gamma z^5 \, dz &= \int_{-1}^{\pi/2} \gamma(t)^5 \gamma'(t) \, dt \\
&= \int_{-1}^{0} (t+1)^5 . 1 \, dt + \int_{0}^{\pi/2} \left(e^{it}\right)^5 i e^{it} \, dt \\
&= \int_{-1}^{0} (t+1)^5 \, dt + \int_{0}^{\pi/2} i e^{5it} \, dt \\
&= \left[\frac{1}{6}(t+1)^6 \right]_{-1}^{0} + \left[\frac{i}{6i} e^{6it} \right]_{0}^{\pi/2} \\
&= \left[\frac{1}{6} - 0 \right] + \frac{1}{6}[-1 - 1] \\
&= -\frac{1}{6}.
\end{aligned}
$$

Remark 5.15

Suppose that $f : D \to \mathbb{C}$ is a complex function such that $f(x)$ is real for x real, for example, $\sin x$. If we take $\gamma : [a, b] \to \mathbb{C}$ given by $\gamma(t) = t$ for $a \leq t \leq b$, then

$$\int_{\gamma} f(z) \, dz = \int_{a}^{b} f(t)\gamma'(t) \, dt = \int_{a}^{b} f(t) \, dt.$$

Thus, by taking a contour along the real line, we can see that contour integration includes the theory of real integration as a special case.

Fundamental Example

The following example will be fundamental to the theory of complex functions.

Example 5.16

For $w \in \mathbb{C}$ define the function $f : \mathbb{C} \backslash \{w\} \to \mathbb{C}$ by $f(z) = (z - w)^n$ where $n \in \mathbb{Z}$. Let C_r be the circular contour with centre w and radius $r > 0$, i.e.,

$$C_r(t) = w + re^{it}, \quad 0 \leq t \leq 2\pi.$$

Then,

$$
\begin{aligned}
\int_{C_r} (z - w)^n \, dz &= \int_{0}^{2\pi} (C_r(t) - w)^n \, C_r'(t) \, dt \\
&= \int_{0}^{2\pi} (w + re^{it} - w)^n \, ire^{it} \, dt \\
&= \int_{0}^{2\pi} r^n e^{int} ire^{it} \, dt \\
&= ir^{n+1} \int_{0}^{2\pi} e^{i(n+1)t} \, dt \\
&= \begin{cases} \dfrac{r^{n+1}}{n+1} \left[e^{i(n+1)t} \right]_0^{2\pi}, & \text{if } n \neq -1, \\[2mm] i \int_0^{2\pi} 1 \, dt, & \text{if } n = -1. \end{cases} \\
&= \begin{cases} 0, & \text{if } n \neq -1, \\ 2\pi i, & \text{if } n = -1. \end{cases}
\end{aligned}
$$

That is,

$$\int_{C_r} \frac{1}{z-w} \, dz = 2\pi i,$$

$$\text{and } \int_{C_r} (z-w)^n \, dz = 0 \text{ for } n \neq -1.$$

Note this result well, this innocuous looking calculation will be used to devastating effect later.

Justification of the Definition of Contour Integral

The definition of contour integral appears to come from nowhere. Why was it chosen as the definition? Why does it contain the derivative of γ? And why, for instance, do we not just define the contour integral $\int_\gamma f(z) \, dz$ to be simply $\int_a^b f(\gamma(t)) \, dt$? The answer to the latter question is the less than satisfying assurance that this does not lead to a good theory.

Nonetheless, we shall show that the definition can be deduced naturally from 'first principles'. We show that it is not some random choice but a one arising from a generalization of the Riemann integral. The following argument is intended to be intuitive and is not entirely rigorous – though it can be made so. (Try this as a not very elementary exercise.)

Let $g : [a, b] \to \mathbb{R}$ be a Riemann integrable function, then $\int_a^b g(f) \, dt$ can be approximated as a sum:

$$\int_a^b g(f) \, dt \approx \sum_{j=1}^n g(t_j)(t_j - t_{j-1})$$

where $a = t_0 < t_1 < t_2 < \cdots < t_{n-1} < t_n = b$. The symbol $\approx$ means approximately. As n get large and $\max(t_j - t_{j-1})$ goes to zero one expects this approximation to improve and in the limit we get the definition of the integral.

If we wanted to generalize the notion of real integral to that of a contour integral, then by analogy $\int_\gamma f(z) \, dz$ should be approximated by the sum

$$\sum_{j=1}^n f(z_j)(z_j - z_{j-1})$$

where z_j are (ordered) points on the image of the contour. These points are ordered in the sense that there exist $t_j \in [a, b]$ such that $z_j = \gamma(t_j)$ and $a = t_0 < t_1 < t_2 < \cdots < t_{n-1} < t_n = b$. We can picture the contour and the line segments $z_j - z_{j-1} = \gamma(t_j) - \gamma(t_{j-1})$ as in the following diagram.

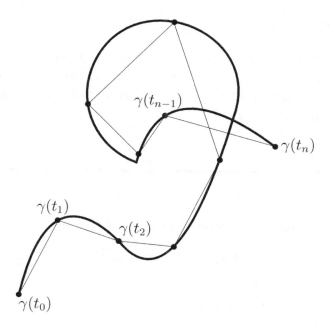

Hence,

$$\int_\gamma f(z)\, dz \approx \sum_{j=1}^{n} f(z_j)(z_j - z_{j-1})$$

$$= \sum_{j=1}^{n} f(\gamma(t_j))\left(\gamma(t_j) - \gamma(t_{j-1})\right)$$

$$= \sum_{j=1}^{n} f(\gamma(t_j))\left(\frac{\gamma(t_j) - \gamma(t_{j-1})}{t_j - t_{j-1}}\right)(t_j - t_{j-1})$$

$$\approx \sum_{j=1}^{n} f(\gamma(t_j))\gamma'(t_j)(t_j - t_{j-1})$$

$$\approx \int_a^b f(\gamma(t))\gamma'(t)\, dt.$$

Therefore, by arguing that, in the limit, the approximation improves we produce a justification for the given definition of contour integration.

Remark 5.17

In the theory of real functions we use the Riemann definition of integral rather than defining the integral as the area under the curve because it is hard to rigorously define area. Note that in the complex case we *have* to use a Riemann integral type approach since it is not even obvious what area under the curve should mean in this case.

This helps explain why we teach the concept of Riemann integration – it can be generalized to other situations whereas the notion of integral as 'area under a curve' cannot. (And, by the way, it is hard to define what we mean by area without first defining integrals!)

Exercises

Exercises 5.18

(i) Sketch the following contours, indicating the direction of travel.

(a) $\gamma_1(t) = (t+1)^2 + i(t+1)$ for $-1 \le t \le 2$;

(b) $\gamma_2(t) = 5e^{2it}$ for $0 \le t \le \pi/2$;

(c) $\gamma_3(t) = 3 + 2e^{-it}$ for $0 \le t \le 4\pi$.

(ii) Let α, β and γ be contours defined as follows:

$$\begin{aligned} \alpha(t) &= t + it^2, & 0 \le t \le 1, \\ \beta(t) &= 1 + e^{it}, & -\pi/2 \le t \le \pi/2, \\ \gamma(t) &= t - it, & 0 \le t \le 1. \end{aligned}$$

Draw a single diagram depicting the contours α, β, γ (including their orientation).

(iii) On a single diagram draw the following contours indicating their orientation:

$$\begin{aligned} \gamma(t) &= t + it, & 0 \le t \le 1/\sqrt{2}, \\ \delta(t) &= e^{it}, & \pi/4 \le t \le 7\pi/4, \\ \epsilon(t) &= \frac{1}{\sqrt{2}}(t - it^2), & 0 \le t \le 1. \end{aligned}$$

(iv) On a single diagram draw the following contours indicating their orientation:

$$\begin{aligned} \alpha(t) &= 2e^{4\pi it}, & 0 \le t \le 1, \\ \beta(t) &= t + i(t-1)^2, & 1 \le t \le 2, \\ \gamma(t) &= -4 + i + 6e^{-2\pi it}, & 2 \le t \le 3. \end{aligned}$$

(v) Sketch and define a contour with image equal to the following sets.

(a) A triangle with vertices 0, 3 and $1 + i$.

(b) A rectangle with vertices $\pm 3 \pm 2i$.

(c) The figure of eight given by $|z - i| = 1$ and $|z - 3i| = 1$ with the top one traversed clockwise and the lower anticlockwise.

(vi) Let ε and R be real numbers with $0 < \varepsilon < R$. Draw the image of $\Gamma = C_R^+ + \gamma^- - C_\varepsilon^+ + \gamma^+$ where C_r^+ is the contour giving the semi-circle in the upper half of the complex plane of radius $r = \varepsilon$ or $r = R$, oriented anticlockwise, γ^- is the straight line contour from $-R$ to $-\varepsilon$ and γ^+ from ε to R. This is an example of **bump contour**. The name is due to ε usually being small giving a bump at the some point.

(vii) Calculate the following contour integrals.

(a) $\int_\gamma \bar{z} \, dz$ where

i. $\gamma(t) = (1 + i)t$, $0 \le t \le 1$,

ii. $\gamma(t) = t + it^2$, $0 \le t \le 1$.

iii. $\gamma(t) = \begin{cases} it, & 0 \le t \le 1, \\ t - 1 + i, & 1 \le t \le 2, \end{cases}$

(b) $\int_\gamma z^2 \, dz$ where

i. $\gamma(t) = t$, $0 \le t \le 1$, (is the answer what you would expect?)

ii. $\gamma(t) = t + it^2$, $0 \le t \le 1$.

(c) $\int_\delta |z| \, dz$ where δ is from exercise (iii) above.

(d) $\int_\gamma |z|^2 \, dz$ where $\gamma(t) = t + it$ for $-1 \le t \le \dfrac{1}{2}$.

(e) $\int_\alpha \operatorname{Im}(z) \, dz$ where

i. $\alpha(t) = it^2 - 3t$ for $-1 \le t \le 3$,

ii. $\alpha(t) = t^2 + i(t + 2)$ for $0 \le t \le 1$,

iii. α is from exercise (ii) above,

iv. α equals β from exercise (iv) above.

(f) $\int_\gamma \operatorname{Re}(z) \, dz$ where

i. γ is from exercise (iii) above,

ii. $\gamma(t) = 2it^2 - 5t$ for $1 \le t \le 2$,

iii. $\gamma(t) = t + 2ti$ for $0 \le t \le 1$,

iv. $\gamma(t) = re^{it}$, where $0 \le t \le 2\pi$ and $r > 0$.

(g) $\int_\gamma \frac{dz}{z}$ where γ describes the semi-circle from -1 to 1 in the upper half of the complex plane.

(h) $\int_\gamma \frac{dz}{z - 3i}$ where

    i. $\gamma(t) = 3i + 4e^{it}$ where $0 \le t \le 2\pi$,

    ii. $\gamma(t) = 3i + 4e^{2\pi it}$ where $0 \le t \le 1$,

    iii. $\gamma(t) = 3i + 4e^{-it}$ where $0 \le t \le 4\pi$.

(viii) Show that, for n and k natural numbers with $k \le n$,

$$\frac{1}{2\pi i} \int_C \frac{(z+1)^n}{z^{k+1}} \, dz = \binom{n}{k}.$$

(ix) Define the **conjugate contour** of the contour γ, denoted $\overline{\gamma}$, to be $\overline{\gamma}(t) = \overline{\gamma(t)}$. Find a formula for

$$\int_{\overline{\gamma}} \overline{f(\overline{z})} \, dz.$$

(x) Show that $f(z) = \overline{z}$ is continuous. Show that $f(z) = |z|$ is continuous.

(xi) Prove or find a counterexample to the statement

$$\text{Re}\left(\int_\gamma f \, dz \right) = \int_\gamma \text{Re}(f) \, dz.$$

(xii) Describe a complex differentiable function $f : \mathbb{C} \to \mathbb{C}$ such that $f(0) = f(1)$ but $f'(z) \ne 0$ for all $z \in \mathbb{C}$. Note that such an example shows that the **mean value theorem** from real analysis does not generalize to complex analysis.

(xiii) Let γ be a contour from α from β. Show that

(a) $\int_\gamma c \, dz = c(\beta - \alpha)$ for all $c \in \mathbb{C}$;

(b) $\int_\gamma z \, dz = \frac{1}{2} \left(\beta^2 - \alpha^2 \right).$

(xiv) Show that

$$\text{Re}\left(\int_\gamma \overline{z} \, dz \right) = \frac{1}{2} \left(|z_1|^2 - |z_0|^2 \right)$$

for all contours γ from z_0 to z_1. (Hint: Let $\gamma(t) = u(t) + iv(t)$.)

Summary

- A contour is a continuous map $\gamma : [a, b] \to \mathbb{C}$ which has continuously differentiable pieces. It is a complex-valued function of a real variable.

- The image of the contour γ is denoted by γ^*.

- Straight line contour from α to β is $\gamma : [0, 1] \to \mathbb{C}$ given by $\gamma(t) = \alpha + t(\beta - \alpha)$.

- Circular contour of radius r based at w is $C_r : [0, 2\pi] \to \mathbb{C}$ given by $C_r(t) = w + re^{it}$.

- The integral of a continuous f along γ is

$$\int_\gamma f = \int_\gamma f(z)\, dz = \int_a^b f(\gamma(t))\gamma'(t)\, dt.$$

- $\displaystyle\int_{C_r} \frac{1}{z - w}\, dz = 2\pi i, \qquad C_r(t) = w + re^{it}, \, 0 \leq t \leq 2\pi.$

- $\displaystyle\int_{C_r} (z - w)^n\, dz = 0$ for $n \neq -1, \qquad C_r(t) = w + re^{it}, \, 0 \leq t \leq 2\pi.$

CHAPTER 6

Properties of Contour Integration

There are a number of well known properties of real integration:

$$\int_a^b \lambda f(x) + \mu g(x)\, dx = \lambda \int_a^b f(x)\, dx + \mu \int_a^b g(x)\, dx, \ \lambda, \mu \in \mathbb{R}, \qquad (6.1)$$

$$\int_a^c f(x)\, dx = \int_a^b f(x)\, dx + \int_b^c f(x)\, dx, \quad a < b < c, \qquad (6.2)$$

$$\int_a^b f(x)\, dx = -\int_b^a f(x)\, dx, \qquad (6.3)$$

$$\int_a^b f(x)\, dx = \int_{\phi^{-1}(a)}^{\phi^{-1}(b)} f(\phi(y))\phi'(y)\, dy, \text{ where } \phi \text{ is a bijection.} \quad (6.4)$$

All of these have analogues in contour integration. We shall now describe them.

In the following the functions will be continuous on some domain D and the contours will be maps into D.

Linearity

Proposition 6.1 (Linearity of contour integrals)
Suppose that $f : D \to \mathbb{C}$ are $g : D \to \mathbb{C}$ are continuous, γ is a contour with image in D and $\lambda, \mu \in \mathbb{C}$. Then (dropping the reference to the integrating variable)

$$\int_\gamma \lambda f + \mu g = \lambda \int_\gamma f + \mu \int_\gamma g.$$

Proof. The proof follows directly from the linearity property of Riemann integration for complex valued functions of a real variable. The details are as follows:

$$\int_\gamma \lambda f + \mu g = \int_a^b (\lambda f + \mu g)(\gamma(t))\gamma'(t)\, dt$$

$$= \int_a^b (\lambda f(\gamma(t)) + \mu g(\gamma(t)))\, \gamma'(t)\, dt$$

$$= \lambda \int_a^b f(\gamma(t))\gamma'(t)\, dt + \mu \int_a^b g(\gamma(t))\gamma'(t)\, dt$$

$$= \lambda \int_\gamma f + \mu \int_\gamma g.$$

$\square$

Contour integration over joins

Now we generalize the result in Equation 6.2. To do this we generalize the notion of the union of two intervals, $[a, b]$ and $[b, c]$.

Definition 6.2
If $\gamma_1 : [a, b] \to \mathbb{C}$ and $\gamma_2 : [b, c] \to \mathbb{C}$ are two contours such that $\gamma_1(b) = \gamma_2(b)$, then their **join** is the contour $\gamma_1 + \gamma_2 : [a, c] \to \mathbb{C}$ given by

$$(\gamma_1 + \gamma_2)(t) = \begin{cases} \gamma_1(t) & a \le t \le b, \\ \gamma_2(t) & b \le t \le c. \end{cases}$$

This is just the precise mathematical way of saying 'do one contour after the other'. We can only define the join if $\gamma_1(b) = \gamma_2(b)$, i.e., the first ends where the second starts.

Exercise 6.3
Prove that the join of two contours is a contour.

Example 6.4
Consider Example 5.4(v). The contour $\alpha : [-1, \pi/2] \to \mathbb{C}$ is given by

$$\alpha(t) = \begin{cases} t + 1, & \text{for } -1 \le t \le 0, \\ e^{it} & \text{for } 0 \le t \le \pi/2. \end{cases}$$

This is a join. We have $\alpha = \gamma_1 + \gamma_2$ where $\gamma_1(t) = t + 1$ for $-1 \le t \le 0$ and $\gamma_2(t) = e^{it}$ for $0 \le t \le \pi/2$.

Remark 6.5

The notation $\gamma_1 + \gamma_2$ may seem wrong as usually the sum of two functions is the pointwise summation. That is, $(\gamma_1 + \gamma_2)(t)$ is defined by $(\gamma_1 + \gamma_2)(t) = \gamma_1(t) + \gamma_2(t)$. However, there are perfectly good reasons for using our 'bad' notation. These become clear in a course on advanced topology but will not be explained here.

Proposition 6.6 (Contour integral of join)

For any continuous function f and contours γ_1 and γ_2, where $\gamma_1 + \gamma_2$ is defined, we have

$$\int_{\gamma_1 + \gamma_2} f = \int_{\gamma_1} f + \int_{\gamma_2} f.$$

Again, this follows from applying the definition and using a Riemann integral property: $\int_a^b = \int_a^c + \int_c^b$. The details are left as an exercise.

Reverse contour

A contour has a direction of travel - we go from one end to the other. What happens if we travel in the opposite direction? In the real case this is Equation 6.3.

Definition 6.7

If $\gamma : [a, b] \to \mathbb{C}$ is a contour, then its **reverse**, denoted $-\gamma$ is the contour $(-\gamma)(t) = \gamma(a + b - t)$, for $a \le t \le b$.

The point is that we do γ backwards. Instead of starting at $\gamma(a)$ we end there, etc.:

$$(-\gamma)(a) = \gamma(a + b - a) = \gamma(b),$$
$$(-\gamma)(b) = \gamma(a + b - b) = \gamma(a).$$

Remark 6.8

Note that again this notation clashes with standard notation for functions. In this case we normally define $-\gamma$ by $(-\gamma)(t) = -\gamma(t)$. This clash is entirely standard and is unproblematic in complex analysis.

Example 6.9

We can reverse the orientation of the contour in the Fundamental Example. That is,

$$
\begin{aligned}
(-C_r)(t) &= C_r(0 + 2\pi - t) \qquad 0 \le t \le 2\pi, \\
&= w + re^{i(2\pi - t)} \\
&= w + re^{-it}e^{2\pi i} \\
&= w + re^{-it}.
\end{aligned}
$$

Thus we get the circle traversed clockwise.

We can now state and prove an analogue of the result $\int_b^a = -\int_a^b$ from real analysis.

Proposition 6.10 (Contour integral of reverse)
For a continuous f and contour γ, we have

$$\int_{-\gamma} f = -\int_{\gamma} f.$$

Proof. Again we apply definitions and use Riemann integral properties, but in this case we also need a simple change of variables:

$$s = a + b - t, \text{ so } dt = -ds.$$

Note that

$$(-\gamma)'(t) = (-1)\gamma'(a + b - t).$$

We have,

$$\begin{aligned}
\int_{-\gamma} f &= \int_a^b f((-\gamma)(t))(-\gamma)'(t) \, dt \\
&= \int_a^b f(\gamma(a + b - t))(-1)\gamma'(a + b - t) \, dt \\
&= -\int_b^a f(\gamma(s))\gamma'(s) \, (-ds) \\
&= \int_b^a f(\gamma(s))\gamma'(s) \, ds \\
&= -\int_a^b f(\gamma(s))\gamma'(s) \, ds \\
&= -\int_{\gamma} f.
\end{aligned}$$

$\square$

Examples 6.11
(i) See Exercise 5.12(ii) and (iii). Here $\gamma_3 = -\gamma_2$ as follows: We have $a = 0$ and $b = 1$, so

$$(-\gamma_2)(t) = \gamma_2(0 + 1 - t) = \gamma_2(1 - t) = (1 - t) + i(1 - t) = \gamma_3(t).$$

This explains why $\int_{\gamma_3} f_3(z) \, dz = -\int_{\gamma_2} f_2(z) \, dz$ in the exercise.

(ii) Consider the contour C_r centred at w. From Proposition 6.10 and the Fundamental Example can see immediately that

$$\int_{-C_r} \frac{1}{z-w}\, dz = -2\pi i.$$

Reparametrisation

Now we shall do the analogue of a change of variables.

Definition 6.12
Let $\gamma : [a,b] \to \mathbb{C}$ be a contour and let $\phi : [c,d] \to [a,b]$ be a function such that

(i) ϕ is continuously differentiable, (that is ϕ is differentiable and ϕ' is continuous),

(ii) $\phi'(s) > 0$ for all $s \in [c,d]$,

(iii) $\phi(c) = a$, and $\phi(d) = b$.

Then the contour $\widetilde{\gamma} : [c,d] \to \mathbb{C}$ defined by $\widetilde{\gamma}(s) = \gamma(\phi(s))$ is called a **reparametrisation** of γ.

Example 6.13
Let $\gamma(t) = e^{it}$ for $0 \leq t \leq 2\pi$. Let $\phi(s) = 2\pi s$ for $0 \leq s \leq 1$. Then $\widetilde{\gamma}(s) = e^{2\pi i s}$ for $0 \leq s \leq 1$.

We think of reparametrised contours as equivalent since we have the following.

Proposition 6.14 (Reparametrized contour integrals are equivalent)
Suppose that $f : D \to \mathbb{C}$ is continuous and γ is a contour. If $\widetilde{\gamma}$ is a reparametrisation of γ, then

$$\int_{\widetilde{\gamma}} f = \int_{\gamma} f.$$

Proof. We have,

$$
\begin{aligned}
\int_{\widetilde{\gamma}} f &= \int_c^d f(\widetilde{\gamma}(s))\, \widetilde{\gamma}'(s)\, ds \\
&= \int_c^d f(\gamma(\phi(s))\, \gamma'(\phi(s))\, \phi'(s)\, ds, \text{ using chain rule,} \\
&= \int_a^b f(\gamma(t))\, \gamma'(t)\, dt, \text{ using } t = \phi(s), \\
&= \int_{\gamma} f.
\end{aligned}
$$

Thus, it does not matter whether we evaluate the integral $\int_\gamma \dfrac{1}{z-w}\, dz$, for example, using $\gamma(t) = e^{it}$ for $0 \le t \le 2\pi$ or $\gamma(t) = e^{2\pi it}$ for $0 \le t \le 1$. (Checking this as an exercise using the definition of contour integral can help illuminate the proof above.)

Remarks 6.15

   (i) If $\gamma_1 : [a, b] \to \mathbb{C}$ and $\gamma_2 : [c, d] \to \mathbb{C}$ are two contours such that $\gamma_1(b) = \gamma_2(c)$, but possibly $b \neq c$, then we can reparametrise γ_2 as

$$\widetilde{\gamma}_2 : [b, d + b - c] \to \mathbb{C} \text{ given by } \widetilde{\gamma}_2(s) = \gamma_2(c - b + s).$$

      Then $\gamma_1(b) = \widetilde{\gamma}_2(b)$ so we can form the join $\gamma_1 + \widetilde{\gamma}_2$. From now on we abuse notation and write this simply as $\gamma_1 + \gamma_2$.

  (ii) If $\widetilde{\gamma}$ is a reparametrisation of γ, then it has the same image as γ. The converse is not true. See exercises.

Putting it all together

Example 6.16
Find the integral $\int_\gamma \dfrac{1}{(z-w)}\, dz$, where γ describes the boundary of the square with corners $w \pm l \pm il$, starting at $w + l - li$ and going anticlockwise.

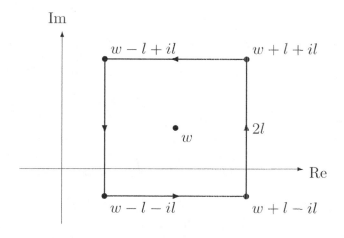

So $\gamma = \gamma_1 + \gamma_2 + \gamma_3 + \gamma_4$. By Proposition 6.6 we have

$$\int_\gamma \frac{dz}{z-w} = \sum_{i=1}^4 \int_{\gamma_i} \frac{dz}{z-w}.$$

We can now compute the individual pieces and by Proposition 6.14 it does not matter how we parametrize the contours.

Consider the following contours: $\nu^\pm(t) = w \pm l + it$, $-l \le t \le l$. Then $\nu^+ = \gamma_1$ and $\nu^- = -\gamma_3$.

We have,

$$
\begin{aligned}
\int_{\nu^\pm} \frac{dz}{z-w} &= \int_{-l}^l \frac{(\nu^\pm)'(t)}{\nu^\pm(t) - w} \, dt \\
&= \int_{-l}^l \frac{i}{\pm l + it} \, dt \\
&= \int_{-l}^l \frac{\pm l - it}{l^2 + t^2} \, i \, dt \\
&= \int_{-l}^l \frac{t}{l^2 + t^2} \, dt \pm i \int_{-l}^l \frac{l}{l^2 + t^2} \, dt \\
&= 0 \pm i \left[\tan^{-1} \frac{t}{l} \right]_{-l}^l \\
&= \pm \frac{i\pi}{2}.
\end{aligned}
$$

Hence,

$$\int_{\gamma_1} \frac{dz}{z-w} = \int_{\nu^+} \frac{dz}{z-w} = \frac{i\pi}{2}.$$

Using Proposition 6.10 we have

$$\int_{\gamma_3} \frac{dz}{z-w} = \int_{-\nu^-} \frac{dz}{z-w} = -\int_{\nu^-} \frac{dz}{z-w} = \frac{i\pi}{2}.$$

Let $\eta^\pm(t) = w + tl \pm il$, $-l \le t \le l$. Then $\eta^+ = -\gamma_2$ and $\eta^- = \gamma_4$. A similar calculation to the one above gives

$$\int_{\gamma_1} = \int_{\gamma_2} = \int_{\gamma_3} = \int_{\gamma_4} = \frac{\pi}{2} i.$$

So,

$$\int_\gamma \frac{dz}{z-w} = \sum_{i=1}^4 \int_{\gamma_i} \frac{dz}{z-w} = 4 \times \frac{\pi}{2} i = 2\pi i.$$

Notice that this coincides with the value of the integral when γ is a circle, see the Fundamental Example, (Example 5.16). This is not a coincidence as we shall see later.

Exercises

Exercises 6.17

(i) Prove Proposition 6.6.

(ii) Show that condition (i), ϕ is continuously differentiable, of the reparametrization is necessary for $\widetilde{\gamma}$ to be a contour.

(iii) Show that the $\phi'(s) > 0$ condition of the reparametrization definition is necessary for Proposition 6.14.

(iv) Consider Remark 6.15(ii). Show that if $\widetilde{\gamma}$ is a reparametrization of γ, then γ and $\widetilde{\gamma}$ have the same image. Give a counterexample to the converse.

(v) Show from the definition of a contour integral that a linear change of coordinates of the domain of the contour does not change the value of the contour integral.

That is, let γ be a contour and $\phi : \mathbb{R} \to \mathbb{R}$ be given by $\phi(t) = \alpha t + \beta$ for $\alpha, \beta \in \mathbb{R}$. For $\widetilde{\gamma} = \gamma \circ \phi$ show that

$$\int_{\widetilde{\gamma}} f = \int_{\gamma} f.$$

(vi) Suppose that $f : D \to \mathbb{C}$ is an **even** function. That is, $f(z) = f(-z)$ for all $z \in D$. Assume that $C_r^* \subset D$ for some $r > 0$ where C_r is the usual circular contour of radius r and centered at the origin. Show that

$$\int_{C_r} f(z)\,dz = 0.$$

Summary

❏ We can join contours.

❏ We can reverse contours.

❑ We can reparametrize contours.

❑ If f, g are continuous on D, γ a contour in D, $\lambda, \mu \in \mathbb{C}$, then

$$\int_\gamma \lambda f + \mu g = \lambda \int_\gamma f + \mu \int_\gamma g.$$

❑ If f is continuous on D, γ_1 and γ_2 contours in D such that the end of γ_1 is the start of γ_2, then

$$\int_{\gamma_1 + \gamma_2} f = \int_{\gamma_1} f + \int_{\gamma_2} f.$$

❑ For f continuous, we have

$$\int_{-\gamma} f = -\int_\gamma f.$$

❑ If f is continuous and $\widetilde{\gamma}$ is a reparametrisation of γ, then

$$\int_{\widetilde{\gamma}} f = \int_\gamma f,$$

CHAPTER 7

The Estimation Lemma

For a continuous real function f, if $|f(x)| \leq M$ for all $x \in [a, b]$, then

$$\left| \int_a^b f(x)\, dx \right| \leq M(b - a).$$

We would like a complex version of this, that is, a bound on $\left| \int_\gamma f(z)\, dz \right|$. The generalization we give is arguably the most useful tool in complex analysis and almost all of the subsequent important proofs in this book use it in one way or another.

The definition of the bound M is straightforward for complex functions: we want M such that $|f(\gamma(t))| \leq M$ for all $t \in [a, b]$. The generalization of $b - a$, the length of the interval $[a, b]$ needs a bit more work and we do this now.

Definition 7.1
Let $\gamma : [a, b] \to \mathbb{C}$ be a contour. The **length of** γ, denoted $L(\gamma)$, is defined to be

$$L(\gamma) = \int_a^b |\gamma'(t)|\, dt.$$

This is a good definition since it agrees with the intuitive notion of the length of a curve: $\gamma'(t)$ is the velocity of a contour, so $|\gamma'(t)|$ is the speed, and the integral of speed with respect to time is distance.

Example 7.2
Let $\gamma(t) = \alpha + t(\beta - \alpha)$ with $0 \leq t \leq 1$ be the straight line contour from α to β.

We have

$$
\begin{aligned}
L(\gamma) &= \int_0^1 |\gamma'(t)|\, dt \\
&= \int_0^1 |\beta - \alpha|\, dt \\
&= |\beta - \alpha| \int_0^1 dt \\
&= |\beta - \alpha|[t]_0^1 \\
&= |\beta - \alpha|.
\end{aligned}
$$

So, the length of the contour from α to β is $|\beta - \alpha|$, which is reassuring.

Exercises 7.3

  (i) Show that the length of the contour given by traversing once round the circle of radius r based at the point $w \in \mathbb{C}$ is $2\pi r$. Generalize this to an arc of a circle subtending an angle θ.

  (ii) Show that $L(-\gamma) = L(\gamma)$.

 (iii) Show that $L(\alpha + \beta) = L(\alpha) + L(\beta)$ where α and β are contours for which we can take the join $\alpha + \beta$.

 (iv) Prove that $L(\gamma) < \infty$. (This is not as obvious as it seems. There exist fractal curves with infinite length. These curves are not allowed in our definition of contour.)

Triangle inequality for complex Riemann integration

We need the following lemma.

Lemma 7.4

Suppose that $g : [a, b] \to \mathbb{C}$ is complex Riemann integrable such that $|g|$ is also Riemann integrable. Then

$$
\left| \int_a^b g(t)\, dt \right| \leq \int_a^b |g(t)|\, dt.
$$

Proof. If the left-hand side is zero, then the statement is trivial. Hence, assume that the left-hand side is non-zero. Define the complex number c by

$$c = \frac{\left| \int_a^b g(t)\, dt \right|}{\int_a^b g(t)\, dt}.$$

We have,

$$
\begin{aligned}
\left| \int_a^b g(t)\, dt \right| &= c \int_a^b g(t)\, dt \\
&= \int_a^b \operatorname{Re}\left(cg(t) \right) dt + i \int_a^b \operatorname{Im}(cg(t))\, dt \\
&= \int_a^b \operatorname{Re}\left(cg(t) \right) dt, \text{ because the left-hand side is real,} \\
&\leq \int_a^b |cg(t)|\, dt, \text{ as } \operatorname{Re}(z) \leq |z|, \\
&= \int_a^b |c|\,|g(t)|\, dt \\
&= \int_a^b |g(t)|\, dt, \text{ as clearly } |c| = 1.
\end{aligned}
$$

$\square$

Estimation Lemma

The next lemma allows us to estimate contour integrals and generalizes the real theorem stated at the start of this chapter.

Lemma 7.5 (Estimation Lemma)
Let $f : D \to \mathbb{C}$ be a continuous complex function and $\gamma : [a, b] \to D$ be a contour. Suppose that $|f(z)| \leq M$ for all $z \in \gamma^*$. Then

$$\left| \int_\gamma f(z)\, dz \right| \leq M\, L(\gamma).$$

Proof. We have,

$$\left| \int_{\gamma} f(z) \, dz \right| = \left| \int_{a}^{b} f(\gamma(t))\gamma'(t) \, dt \right|$$

$$\leq \int_{a}^{b} |f(\gamma(t))\gamma'(t)| \, dt \text{ by Lemma 7.4,}$$

$$= \int_{a}^{b} |f(\gamma(t))| \, |\gamma'(t)| \, dt$$

$$\leq \int_{a}^{b} M|\gamma'(t)| dt$$

$$= M \int_{a}^{b} |\gamma'(t)| dt$$

$$= ML(\gamma).$$

$\square$

We can give a name to the M appearing in the theorem.

Definition 7.6

Let $f : D \to \mathbb{C}$ be a complex function and let $S \subseteq D$ be a subset of D. If $|f(z)| \leq M$ for all $z \in S$, then M is called an **upper bound** for the function f on S. If $M \leq |f(z)|$ for all $z \in S$, then M is called a **lower bound** for the function f on S.

In either case we can refer to M as a **bound**.

Example 7.7

Show that

$$\left| \int_{\gamma_R} \frac{e^z}{z^2 + 1} \, dz \right| \leq \frac{\pi R}{R^2 - 1}$$

where γ_R describes the semi-circle from iR to $-iR$ in the left half-plane $\{z : \text{Re}(z) \leq 0\}$ with $R > 1$. This contour is shown in the next figure.

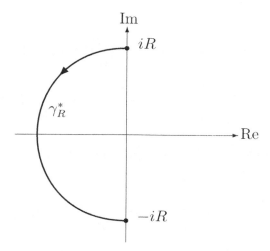

Solution: If $z = x + iy$ lies on the image of γ_R, then $x \leq 0$ and so

$$|e^z| = e^{\mathrm{Re}(z)} = e^x \leq e^0 = 1.$$

Also, when z lies on γ_R^*, then $|z| = R$, so (using the reverse triangle inequality)

$$|z^2 + 1| \geq \left||z^2| - 1\right| = \left||z|^2 - 1\right| = |R^2 - 1| = R^2 - 1$$

(The last equality follows from $R > 1$.) Thus $|z^2 + 1| \geq R^2 - 1$. (That is, $R^2 - 1$ is a lower bound for $|z^2 + 1|$ on the contour.) Hence,

$$\left|\frac{e^z}{z^2 + 1}\right| \leq \frac{1}{R^2 - 1}, \text{ for all } z \in \gamma_R^*.$$

We know that $L(\gamma_R) = \pi R$ as γ_R^* is semi-circle of radius R, so by the Estimation Lemma,

$$\left|\int_{\gamma_R} \frac{e^z}{z^2 + 1}\, dz\right| \leq \frac{1}{R^2 - 1} L(\gamma_R) = \frac{\pi R}{R^2 - 1}.$$

Example 7.8
Let $\gamma(t) = Re^{it}$, $0 \leq t \leq \pi$ describe a semi-circle, with $R > \sqrt[3]{2}$.
   Show that

$$\left|\int_\gamma \frac{e^{2iz}}{(z^3 - 2)^2}\, dz\right| \leq \frac{\pi R}{(R^3 - 2)^2}.$$

Solution: For $z \in \gamma^*$ we have $|z| = R$ and so

$$|(z^3 - 2)^2| = |z^3 - 2|^2 \geq \left||z^3| - |-2|\right|^2 = |R^3 - 2|^2 = (R^3 - 2)^2.$$

Thus, on the contour

$$\frac{1}{|(z^3 - 2)^2|} \leq \frac{1}{(R^3 - 2)^2}.$$

For the numerator of the integrand:

$$|e^{2iz}| = |e^{iz}|^2 = |e^{\text{Re}(iz)}|^2 = |e^{-\text{Im}(z)}|^2.$$

But $\text{Im}(z) \geq 0$ because γ^* lies in the upper half-plane. Thus, $0 < e^{-\text{Im}(z)} \leq 1$ and so $|e^{2iz}| \leq 1^2 = 1$.

Hence,

$$\left| \frac{e^{2iz}}{(z^3 - 2)^2} \right| \leq \frac{1}{(R^3 - 2)^2}.$$

The length of γ is πR since it gives a semi-circle. Thus, the inequality follows from the Estimation Lemma.

Exercises 7.9

(i)   (a) Find a bound for $|f(z)| = \dfrac{1}{|z^3 + 3|}$ on $\gamma(t) = 5e^{it}$ for $0 \leq t \leq \pi/2$.

     (b) Find the length of γ.

     (c) Estimate the contour integral $\displaystyle\int_\gamma \frac{1}{|z^3 + 3|}\, dz$.

(ii) Let γ be the straight line contour from $1 + 3i$ to $-2 - i$.

     (a) Find an upper bound for $|e^z|$ on γ. (Hint: Use Theorem 3.1.)

     (b) Find an upper bound for $|z|$ on γ. (Hint. Draw a picture of γ^*.)

     (c) Find the length of γ. (No integrating necessary!)

     (d) Find an estimate for the integral $\displaystyle\int_\gamma z e^z\, dz$.

Remarks 7.10

(i) The constant M always exists: On the image of γ the function $|f(z)|$ will always be bounded because the map $t \mapsto |f(\gamma(t))|$ is a continuous real function on $[a, b]$, and hence is bounded. So $M = \sup_{z \in \gamma^*}\{|f(z)|\}$ will do as a bound. Anything bigger than this can also be used as a bound.

(ii) Although the estimation lemma looks rather crude it will be extremely useful for us again and again. In practice we will show that some integral or function is zero by showing that $ML(\gamma)$ tends to zero for a family of contours or functions.

For instance, consider Example 7.7. As $R \to \infty$ we have $\dfrac{\pi R}{R^2 - 1} \to 0$ and hence

$$\lim_{R \to \infty} \int_{\gamma_R} \frac{e^z}{z^2 + 1} \, dz = 0.$$

(Here we use $\lim_{R \to \infty} |f(R)| = 0$ implies $\lim_{R \to \infty} f(R) = 0$. The proof of this is an exercise.)

Exercises

Exercises 7.11

(i) Consider the assumptions of Lemma 7.4. Is the assumption that $|g|$ is Riemann integrable necessary when g is complex Riemann integrable?

(ii) Suppose that $\gamma : [a, b] \to \mathbb{C}$ is a contour. Show that $L(\gamma) \geq |\gamma(b) - \gamma(a)|$. What does this mean geometrically?

(iii) Calculate the lengths of

   (a) $\alpha : [1, 3] \to \mathbb{C}$ given by $\alpha(t) = 3 - t + i(t + 5)$,

   (b) $\beta : [0, 1] \to \mathbb{C}$ given by $\beta(t) = t + 3it^2$. (The integral in this case is not straightforward and you will probably need to look it up.)

(iv) Show that $|\sin z| \leq \dfrac{1}{2} \left(e^{-\mathrm{Im}(z)} + e^{\mathrm{Im}(z)} \right)$ for all $z \in \mathbb{C}$. Hence deduce that $|\sin x| \leq 1$ for all $x \in \mathbb{R}$.

(v) Find a bound for $\displaystyle\int_{\gamma} (z + 2)e^{iz} \, dz$ where γ is the straight line from $\alpha \in \mathbb{C}$ to $\beta \in \mathbb{C}$.

(vi) Prove that $|\sinh z| \leq e^{|z|}$ for all $z \in \mathbb{C}$.

(vii) Estimate $\left| \int_{\gamma} z^2 \, dz \right|$ for $\gamma(t) = e^{it}$, $0 \leq t \leq \pi/2$. How does this compare with the true value calculated in Exercise 5.12.

(viii) Find an estimate for $\left| \dfrac{e^z}{z} \right|$, where $z = e^{i\theta}$, $\theta \in \mathbb{R}$. (I.e., find an M such that $\left| \dfrac{e^z}{z} \right| \leq M$, where $z = e^{i\theta}$, $\theta \in \mathbb{R}$.)

(ix) Set $\gamma(t) = 3e^{it}$ $(0 \le t \le \pi)$. Use the Estimation Lemma to show that

$$\left| \int_\gamma \frac{e^z}{z-1} \, dz \right| \le \frac{3}{2} \pi e^3.$$

(x) Let $\gamma(t)$ describe the semi-circle Re^{it}, where $0 \le t \le \pi$, and $R > 3$. Show that

$$\left| \int_\gamma \frac{e^{3iz}}{(z^2+4)(z^2+9)} \, dz \right| \le \frac{\pi R}{(R^2-4)(R^2-9)}.$$

(xi) Apply the Estimation Lemma to estimate the integral $\int_\gamma e^{1/z} \, dz$ where $\gamma(t) = 7e^{it}$, $0 \le t \le \pi/2$.

(xii) Suppose that $f : \mathbb{C} \to \mathbb{C}$ is a continuous function such that for all

$$|f(z)| \le \frac{1}{R^2} \text{ for } |z| = R.$$

Prove that

$$\lim_{R \to \infty} \int_{\gamma_R} f(z) \, dz = 0$$

for $\gamma_R(t) = Re^{it}$ with $0 \le t \le \pi$.

(xiii) Let γ_r be the arc of a circle $|z| = r$, $r > 1$ of angle $\pi/5$ in the upper half of the complex plane. Use the Estimation Lemma to show that

$$\lim_{r \to \infty} \int_{\gamma_r} \frac{ze^{20iz}}{z^{20}+1} \, dz = 0.$$

(xiv) Use the Estimation Lemma to show that

$$\int_{\gamma_R} \frac{ze^{-4iz}}{z^3+5} \, dz \le \frac{\pi R^2}{R^3-5}$$

where γ_R is the arc of a circle $|z| = R$ of angle $\pi/4$ in the lower half of the complex plane and $R > \sqrt[3]{5}$.

Summary

❏ The length of γ, denoted $L(\gamma)$, is defined to be

$$L(\gamma) = \int_a^b |\gamma'(t)| \, dt.$$

❑ Suppose that f is continuous and $|f(z)| \leq M$ for all $z \in \gamma^*$. Then

$$\left| \int_\gamma f(z)\, dz \right| \leq M\, L(\gamma).$$

❑ $M = \sup\limits_{z \in \gamma^*}\{|f(z)|\}$ will do as a bound for $|f(z)|$ on γ^*.

CHAPTER 8

Complex Differentiation

We turn now to the theory of complex differentiation. In contrast to integration this is very similar to differentiation in real analysis and, in fact, most of the initial results are identical to the real case. However, we see in later chapters that the assumption of differentiability in complex analysis has profound consequences.

For the moment, the new and interesting element arises with the Cauchy-Riemann equations. This is a pair of equations which link the derivatives of the real and imaginary parts of a complex differentiable function. Surprisingly, these equations can be used in the study of an important differential equation.

Complex Differentiation

Definition 8.1
Let $f : D \to \mathbb{C}$ be a complex function. Then, f is **complex differentiable at** w if

$$\lim_{h \to 0} \frac{f(w+h) - f(w)}{h}, \; h \in \mathbb{C}$$

exists. We write $f'(w)$ for this limit, or $\dfrac{df}{dz}(w)$.

We say that f is **complex differentiable on** D if it is complex differentiable at w for all $w \in D$. If it is obvious that we are dealing with a complex function, then we shall just say f is **differentiable**.

The definition looks the same as in the real case, but the fact that h can go to zero from *any* direction in the complex plane makes a huge difference; it imposes strong restrictions.

Remarks 8.2

(i) Complex differentiable functions are also called **holomorphic** or **analytic**. Note that in the theory of real functions analytic is different to differentiable. Analytic means that the function is equal to its Taylor series. The reason why there is no distinction in the complex case will be revealed in Chapter 13.

(ii) As in the real case, the definition of differentiability at w is equivalent to the existence of

$$\lim_{z \to w} \frac{f(z) - f(w)}{z - w}.$$

Example 8.3

The function $f : \mathbb{C} \to \mathbb{C}$ given by $f(z) = z^2$ is differentiable on $\mathbb{C}$.

Let $w \in \mathbb{C}$. Then

$$
\begin{aligned}
\lim_{h \to 0} \frac{(w+h)^2 - w^2}{h} &= \lim_{h \to 0} \frac{w^2 + 2wh + h^2 - w^2}{h} \\
&= \lim_{h \to 0} 2w + h \\
&= 2w.
\end{aligned}
$$

That is, $f'(w) = 2w$, as expected.

Of course, mathematicians do not want to do anything as clumsy or ugly as using first principles. We would use a theorem such as the following.

Proposition 8.4

Let f and g be complex functions on the domain D.

(i) If f and g are differentiable at $w \in D$, then so are

  (a) $f + g$, and $(f+g)'(w) = f'(w) + g'(w)$;

  (b) fg, and $(fg)'(w) = f(w)g'(w) + f'(w)g(w)$;

  (c) f/g, and $(f/g)'(w) = \dfrac{g(w)f'(w) - f(w)g'(w)}{g(w)^2}$ provided $g(w) \neq 0$.

(ii) **Chain Rule.** Suppose $f : D \to \mathbb{C}$ and $g : E \to \mathbb{C}$ are complex functions with $f(D) \subseteq E$. If f is differentiable at w and g is differentiable at $f(w)$, then $g \circ f$ is differentiable at w with $(g \circ f)'(w) = g'(f(w))f'(w)$.

Proof. These are proved in the same way as the real case. $\qquad \square$

Example 8.5

If $f(z) = \alpha z^n$ for some complex constant α and positive integer n, then $f'(z) = n\alpha z^{n-1}$. The proof can be done by induction on n using the product rule and the initial case $n = 1$.

We can then use this result and the proposition to show that all complex polynomials are complex differentiable.

Warning! 8.6

Unlike continuity, complex differentiablity isn't the same as being differentiable with respect to two real variables. There is a connection as we see shortly. In fact it is this difference that makes complex analysis so different to real analysis.

Exercise 8.7

Show that if $f : D \to \mathbb{C}$ is differentiable at c, then f is continuous at c.

We don't yet know that we can differentiate series term-by-term and so can't immediately prove that e^z, $\cos z$ and $\sin z$ are differentiable, or that their derivatives are what we expect them to be. It is possible to find their derivatives from first principles, (you can try this as an exercise if you are keen), but we will delay a proof till later and just state the following.

Theorem 8.8

The elementary functions have the expected derivatives:

(i) $\dfrac{d}{dz} e^z = e^z$, *for all* $z \in \mathbb{C}$.

(ii) $\dfrac{d}{dz} \sin z = \cos z$ *for all* $z \in \mathbb{C}$.

(iii) $\dfrac{d}{dz} \cos z = -\sin z$ *for all* $z \in \mathbb{C}$.

Fundamental Theorem of Calculus

Every good theory of integration and differentiation should have an analogue of the Fundamental Theorem of Calculus. Let's now state and prove one for complex integration and differentiation.

Theorem 8.9 (Fundamental Theorem of Calculus for Contour Integrals)

Let $f : D \to \mathbb{C}$ be a continuous complex function and $\gamma : [a, b] \to D$ be a contour. Suppose that there exists a complex differentiable $F : D \to \mathbb{C}$ such that $F' = f$.

Then,

$$\int_\gamma f(z)\,dz = F(\gamma(b)) - F(\gamma(a)).$$

Proof. Let $a = a_0 < a_1 < \cdots < a_n = b$ be a dissection of $[a, b]$ such that $\gamma'|[a_{j-1}, a_j]$ is continuous for all j. Then,

$$
\begin{aligned}
\int_\gamma f(z)\,dz &= \int_\gamma F'(z)\,dz \\
&= \int_a^b F'(\gamma(t))\gamma'(t)\,dt \\
&= \sum_{j=1}^n \int_{a_{j-1}}^{a_j} F'(\gamma(t))\gamma'(t)\,dt \\
&= \sum_{j=1}^n \int_{a_{j-1}}^{a_j} (F \circ \gamma)'(t)\,dt \\
&= \sum_{j=1}^n [(F \circ \gamma)(t)]_{a_{j-1}}^{a_j}, \quad \text{by Proposition 4.10,} \\
&= \sum_{j=1}^n (F(\gamma(a_j)) - F(\gamma(a_{j-1}))) \\
&= F(\gamma(b)) - F(\gamma(a)).
\end{aligned}
$$

$\square$

Example 8.10
Consider Example 5.11. Then, $f(z) = z^2$, $\gamma(0) = 2$ and $\gamma(1) = 2 + i$.

Obviously, $F(z) = \frac{1}{3}z^3$ is such that $F'(z) = f(z)$. Then, by the Fundamental Theorem of Calculus for contour integrals,

$$
\begin{aligned}
\int_\gamma f(z)\,dz &= \frac{1}{3}\left((\gamma(1))^3 - (\gamma(0))^3\right) \\
&= \frac{1}{3}\left((2+i)^3 - 2^3\right) \\
&= -2 + \frac{11}{3}i.
\end{aligned}
$$

The function F in the theorem is called an **antiderivative** for f. (Such a function is also called a **primitive**.) How common are antiderivatives for continuous complex functions? Do they always exist? For real continuous functions we know that the Riemann integral of a function can be found and this will be an antiderivative. Unfortunately, the analogue is not true for continuous complex functions, not even differentiable ones. Consider this corollary of the Fundamental Theorem of Calculus and the following example.

Corollary 8.11

With the assumptions of the above theorem suppose that γ is any closed contour. Then

$$\int_\gamma f(z)\,dz = 0.$$

Proof. The definition of a closed contour is that $\gamma(a) = \gamma(b)$. So,

$$\int_\gamma f(z)\,dz = F(\gamma(b)) - F(\gamma(b)) = 0.$$

$\square$

Example 8.12

Let $f(z) = 1/z$. Then, f is differentiable on $D = \mathbb{C}\backslash\{0\}$. Let γ be the unit circle round the origin, traversed once anti-clockwise. Then, by the fundamental example we know

$$\int_\gamma f(z)\,dz = \int_\gamma \frac{1}{z}\,dz = 2\pi i.$$

Therefore, if there existed an $F : D \to \mathbb{C}$ such that $F' = f$, then the corollary would be contradicted. Hence, antiderivatives do not always exist.

We can now prove a result we would expect by analogy with real analysis: a function with derivative equal to zero everywhere is constant.

First we need a definition.

Definition 8.13

A domain D is called **path-connected** if for each pair $\alpha, \beta \in D$, there exists a contour $\gamma : [a, b] \to D$ such that $\gamma(a) = \alpha$ and $\gamma(b) = \beta$.

Intuitively speaking, a domain is path-connected if we can draw a curve from any point to any other.

Examples 8.14

(i) The domains $D = \mathbb{C}$ and $D = \mathbb{C}\backslash\{0\}$ are both path-connected.

(ii) The domain $D = \mathbb{C}\backslash\{z \mid z \text{ is real}\}$ is not path-connected. As contours are continuous maps we cannot construct one starting below and finishing above the real line, without crossing that line. However, the real line is not in D.

The next theorem is not too surprising.

Theorem 8.15
Suppose that D is a path-connected domain in $\mathbb{C}$ and that $f : D \to \mathbb{C}$ is differentiable such that $f'(z) = 0$ for all $z \in D$. Then, f is constant.

Proof. Take any α and β in D. Then, as D is path-connected, there exists a path $\gamma : [a, b] \to D$, such that $\gamma(a) = \alpha$ and $\gamma(b) = \beta$. By the Fundamental Theorem of Calculus,

$$f(\beta) - f(\alpha) = f(\gamma(b)) - f(\gamma(a)) = \int_\gamma f'(z)\,dz = \int_\gamma 0\,dz = 0.$$

Thus $f(\beta) = f(\alpha)$. Since these were general points of D, we deduce that f is constant. $\square$

Cauchy-Riemann Equations

If f is a differentiable complex function, then the real and imaginary parts are differentiable as real functions. An interesting fact is that the derivatives are closely related.

Theorem 8.16 (Cauchy-Riemann equations)
Suppose that f is a complex function with $f(x + iy) = u(x, y) + iv(x, y)$. If f is complex differentiable at $z = x + iy$, then the partial derivatives for u and v exist and

$$u_x = v_y \text{ and } v_x = -u_y \text{ at } (x, y)$$

where u_x is the partial derivative of u with respect to x, etc.

The two equations in the theorem are called the **Cauchy-Riemann equations** after two of the founders of complex analysis.

Proof. As f is differentiable we have

$$f'(z) = \lim_{h \to 0} \frac{f(z + h) - f(z)}{h}$$

where $h \in \mathbb{C}$. Let $h \to 0$ through real values. Then

$$f'(z) = \lim_{h \to 0} \frac{u(x + h, y) + iv(x + h, y) - (u(x, y) + iv(x, y))}{h}$$
$$= \lim_{h \to 0} \frac{u(x + h, y) - u(x, y)}{h} + i\frac{v(x + h, y) - v(x, y)}{h}$$
$$= u_x(x, y) + iv_x(x, y).$$

Thus $u_x(x, y)$ exists as it is equal to $\text{Re}(f'(z))$. Similarly v_x exists at (x, y).

Next we can let $h \to 0$ through *imaginary* values, so let $h = il$ where $l \in \mathbb{R}$:

$$f'(z) = \lim_{il \to 0} \frac{u(x, y+l) + iv(x, y+l) - (u(x, y) + iv(x, y))}{il}$$

$$= \lim_{l \to 0} \frac{1}{i} \left(\frac{u(x, y+l) - u(x, y)}{l} + i \frac{v(x, y+l) - v(x, y)}{l} \right)$$

$$= -iu_y(x, y) + v_y(x, y).$$

Hence, the partial derivatives with respect to y exist also. The equations follow from equating the real and imaginary parts arising in

$$u_x(x, y) + iv_x(x, y) = f'(z) = -iu_y(x, y) + v_y(x, y).$$

$\square$

Note that the corollary says that if f is differentiable, then the equations hold, but says nothing of the converse, which is not true anyway as we shall see!

Examples 8.17

(i) Let $f(z) = z^2 = (x + iy)^2$. Then, $u(x, y) = x^2 - y^2$ and $v(x, y) = 2xy$. We see that

$$u_x = 2x = v_y \qquad \text{and} \qquad u_y = -2y = -v_x,$$

so the Cauchy-Riemann equations hold on all of $\mathbb{C}$

(ii) Let $f(z) = |z|^2 = x^2 + y^2$. Then $u = x^2 + y^2$ and $v = 0$. Thus,

$$\begin{aligned} u_x &= 2x, & v_y &= 0, \\ u_y &= 2y, & v_x &= 0. \end{aligned}$$

The Cauchy-Riemann equations are satisfied for $2x = 0$ and $2y = -0$, that is for $(x, y) = (0, 0)$. Therefore f is not differentiable at z for $z \in \mathbb{C} \backslash \{0\}$. We can check whether f is differentiable at the origin using the definition:

$$\frac{f(0 + h) - f(0)}{h} = \frac{|h|^2 - 0}{h} = \frac{h\bar{h}}{h} = \bar{h} \to 0 \text{ as } h \to 0,$$

so $f'(0) = 0$. Therefore, f is complex differentiable only at the origin.

Exercise 8.18

Show that the function $f(z) = |z|$ is not differentiable anywhere in $\mathbb{C}$. (Note that it is continuous everywhere!)

Remark 8.19

The preceding exercise shows that complex differentiability is imposing a stronger condition than real differentiability, because for real points of $\mathbb{C}$ the function $f(x) = |x|$ *is* differentiable, except at $x = 0$.

The condition is stronger because we require $f'(w)$ to exist as $z \to w$ from *all* directions, not just real ones.

Remark 8.20

The converse to Theorem 8.16 is false. For example, let

$$f(z) = \begin{cases} 0, & \text{if } x = 0 \text{ or } y = 0, \\ 1, & \text{otherwise.} \end{cases}$$

Then, it is easy to check that the equations are satisfied at $(x, y) = (0, 0)$. (Since along the x and y axes the function is constant.) However, if f were differentiable at $(0, 0)$, then it would be continuous there which it clearly is not.

Nonetheless, we have the following theorem as a partial converse to Theorem 8.16.

Theorem 8.21

Let U be an open set in $\mathbb{C}$ and $f : U \to C$ be a complex function. Suppose that u_x, u_y, v_x, v_y are continuous at $(x, y) \in U$, and satisfy the Cauchy-Riemann equations. Then f is complex differentiable at $x + iy$.

The theorem will not be proved as it will only be used in exercises but it is useful to know that the statement is true.

How to remember the equations

There is an easy way to remember the Cauchy-Riemann equations and that is to deduce them from the differentiability of $f(z) = z$. Most students can remember that there are two equations involving the four possible first derivatives for u and v, i.e., u_x, v_x, u_y and v_y. The difficulty is in deciding which of these derivatives are equal and which are the 'negatives' of each other.

For $f(z) = z = x + iy$ we have $u(x, y) = x$ and $v(x, y) = y$ and hence

$$\begin{aligned} u_x &= 1, & v_x &= 0, \\ u_y &= 0, & v_y &= 1. \end{aligned}$$

It should be clear that although there are two possibilities for equalities there is only one choice for 'negatives', i.e., $0 = -0$. So, in going from this particular example to the general we must have $u_y = -v_x$. This leaves the only other possibility to be $u_x = v_y$.

Harmonic functions

It turns out that the real and imaginary parts of a twice differentiable complex function satisfy an important equation that appears in many applications.

Definition 8.22
Let φ be a twice-differentiable function of two real variables on an open set. Then φ is a **harmonic function** if it satisfies Laplace's equation:

$$\varphi_{xx} + \varphi_{yy} = 0.$$

Laplace's equation is important in potential theory, electrostatics, fluid flow, heat flow, and many other areas. Since harmonic functions are solutions of Laplace's equation they are of immense interest.

Example 8.23
Let $\varphi(x, y) = x^2 - y^2$. Then $\varphi_{xx} = 2$ and $\varphi_{yy} = -2$, so $\varphi_{xx} + \varphi_{yy} = 0$.

Example 8.24
Let $\varphi : \mathbb{R}^2 \backslash \{0, 0\} \to \mathbb{R}$ be defined by $\varphi(x, y) = \dfrac{x}{x^2 + y^2}$. Then,

$$\varphi_{xx} + \varphi_{yy} = \frac{2x(x^2 - 3y^2)}{(x^2 + y^2)^3} + \frac{2y(3x^2 - y^2)}{(x^2 + y^2)^3} = 0.$$

Complex analysis provides many examples of harmonic functions, as the next theorem shows.

Theorem 8.25
If $f : D \to \mathbb{C}$ is complex differentiable of the form $f(x + iy) = u(x, y) + iv(x, y)$ and u and v are twice differentiable functions, then u and v are both harmonic:

$$u_{xx} + u_{yy} = 0 \text{ and } v_{xx} + v_{yy} = 0.$$

Proof. We use the Cauchy-Riemann equations:

$$u_{xx} + u_{yy} = (u_x)_x + (u_y)_y = (v_y)_x + (-v_x)_y = v_{xy} - v_{yx} = 0.$$

Similarly for v.   □

Remark 8.26
This theorem says that we can produce solutions to an important differential equation. This is a large supply of solutions since there is a large supply of complex differentiable functions, for example polynomials and rational functions, i.e., functions of the form $f(z) = p(z)/q(z)$ where p and q are polynomials.

Not only that, each complex differentiable function produces two solutions to the differential equation.

We can in some sense reverse this process. First a definition.

Definition 8.27

A function v is called a **harmonic conjugate** of u if u is harmonic and f defined by $f(x + iy) = u(x, y) + iv(x, y)$ is complex differentiable.

Given an harmonic function u there exists a harmonic conjugate v (at least on some open set). We shall show how to find harmonic conjugates by example. You can fashion a proof from the ideas in the example.

Example 8.28

Let $u(x, y) = 2xy$, then u is harmonic. (Exercise!) We can construct v using the Cauchy-Riemann equations. Because

$$2y = u_x = v_y \implies v = \int 2y \, dy = y^2 + g(x) \text{ where } g \text{ is a function of } x,$$

and

$$2x = u_y = -v_x \implies v = \int -2x \, dx = -x^2 + h(y) \text{ where } h \text{ is a function of } y,$$

we can deduce that $v = y^2 - x^2 + C$ where C is a constant.

Note that the presence of the constant shows that the conjugate need not be unique.

A clear victory for complex analysis

The Cauchy-Riemann equations gives us our first victory for abstracting from real to complex analysis. By studying complex differentiable functions we produce a huge supply of solutions to an important real problem. Not only that each complex function gives two distinct real solutions. The important lesson here is that complex provides an insight into real.

Exercises

Exercises 8.29

(i) Show that $f(z) = \overline{z}$ is not differentiable for $z \neq 0$. Is it differentiable at 0? Give reasons.

(ii) Show that $f(z) = \text{Re}(z)$ and $f(z) = \text{Im}(z)$ are not differentiable for all $z \in \mathbb{C}$.

(iii) Find the derivatives of the following functions:

  (a) $\sinh(z)$, $\cosh(z)$, and $\tanh(z)$ (the latter is the **hyperbolic tangent** defined by $\sinh(z)/\cosh(z)$),

  (b) $\dfrac{az + b}{cz + d}$ where $a, b, c, d \in \mathbb{C}$,

(iv) Where are the following funtions *not* differentiable?

$$\text{(a)} \quad |z + 1|^2, \qquad \text{(b)} \quad \frac{1}{z^2}, \qquad \text{(c)} \quad \tan z, \qquad \text{(d)} \quad \frac{\sin^2 z}{z(z^2 + 1)}.$$

(v) Let $f(z) = a_n(z - z_0)^n + a_{n-1}(z - z_0)^{n-1} + \cdots + a_1(z - z_0) + a_0$. Show that $a_k = \dfrac{f^{(k)}(z_0)}{k!}$. Hence write $z^3 + 3z^2 + 11$ as a polynomial in $z - 2i$.

(vi) Prove Theorem 8.8.

(vii) $\int_\gamma \sin(z) \cos^3(z) \, dz$, where $\gamma(t) = e^{it}$ for $0 \le t \le \pi/2$.

(viii) Suppose that $f : D \to \mathbb{C}$ is injective. Show that if f^{-1} denotes the inverse of f on its image, then for all $z \in f(D)$

$$\frac{df^{-1}}{dz}(z) = \frac{1}{f'(z)}.$$

(ix) Show that $\dfrac{d}{dz} \text{Log } z = \dfrac{1}{z}$ for $z \in \mathbb{C} \backslash \{\text{non-positive real numbers}\}$. (Hint: Use the chain rule.) Why does this not contradict Example 8.12?

(x) Let D be a path-connected open set and $f : D \to \mathbb{C}$ be a continuous function. Show that if f has an antiderivative, then this antiderivative is unique up to an additive constant. (That is, if f has an antiderivative, then there exists an antiderivative F such that if G is another antiderivative for f, then $F - G$ is a constant function.)

(xi) Let p be a polynomial and let γ be a closed contour. Prove that

$$\int_\gamma p(z) \, dz = 0.$$

(xii) Suppose that D is an open set in $\mathbb{C}$, $\gamma : [a, b] \to D$ a contour, $f : D \to \mathbb{C}$ and $g : D \to \mathbb{C}$ are differentiable with continuous derivatives. Show that **integration by parts** holds. That is,

$$\int_\gamma f'g = f(\gamma(b))g(\gamma(b)) - f(\gamma(a))g(\gamma(a)) - \int_\gamma fg'.$$

Find

  (a) $\int_\gamma z \sin z \, dz$, where $\gamma(t) = t$, $0 \le t \le \pi$,

  (b) $\int_\gamma z^2 \sin z \, dz$ where $\gamma(t) = Re^{it}$, $0 \le t \le \pi$,

(xiii) Show that $f(x + iy) = \sqrt{|xy|}$ is not differentiable at 0 but satisfies the Cauchy-Riemann equations there.

(xiv) Consider $\varphi : \mathbb{R}^2 \backslash \{0, 0\} \to \mathbb{R}$ defined by $\varphi(x, y) = \dfrac{x}{x^2 + y^2}$ from Example 8.24. Find a harmonic conjugate for φ. Find a complex analytic function f on $\mathbb{C} \backslash \{0\}$ such that $\mathrm{Re}(f) = \varphi$.

(xv) Show that $f(z) = \overline{z}$ is not differentiable for $z \ne 0$. Is f differentiable at 0? Give reasons.

(xvi) Find a harmonic conjugate for $u(x, y) = \sin x \cosh y$.

(xvii) Find the harmonic conjugate pairs that arise from $f(z) = z^n$ for $n = 1, 2, 3, 4$.

(xviii) Consider the complex function $f(z) = \dfrac{z - 1}{z + 1}$. Without differentiating show that the function $\varphi(x, y) = \dfrac{x^2 + y^2 - 1}{(x + 1)^2 + y^2}$ is harmonic. Find a harmonic conjugate.

(xix) Show that $\phi(x, y) = \dfrac{1}{2} \ln(x^2 + y^2)$ is harmonic for $x > 0$ and find a harmonic conjugate.

(xx) Show that in polar coordinates the Cauchy-Riemann equations are

$$r \frac{\partial u}{\partial r} = \frac{\partial v}{\partial \theta}, \qquad \frac{\partial u}{\partial \theta} = -r \frac{\partial v}{\partial r}.$$

(xxi) Let D is a path-connected domain in $\mathbb{C}$ and $f : D \to \mathbb{C}$ be differentiable. Show that each one of the following implies that f is constant:

(a) f is real,

(b) $|f|$ is constant,

(c) $\text{Arg}(f)$ is constant.

[Hint: For (b) and (c) use the Cauchy-Riemann equations.]

(xxii) Let u and v be harmonic functions. Show that $\lambda u + \mu v$ is harmonic for all $\lambda, \mu \in \mathbb{C}$ and show that uv and $u^2 - v^2$ harmonic if u and v are harmonic conjugates.

(xxiii) (Existence of antiderivatives I.) Let $f : D \to \mathbb{C}$ be a continuous complex function on the path-connected domain D. Suppose that $\int_\gamma f(z)\,dz = 0$ for all closed γ in D.

Fix $z_0 \in D$. Define $F : D \to \mathbb{C}$ by

$$F(z) = \int_\gamma f(\zeta)\,d\zeta$$

where γ is any contour from z_0 to z.

Prove that

(a) F is well-defined, i.e., does not depend on the choice of γ;

(b) F is an antiderivative for f.

Summary

❑ The complex function f is complex differentiable at w if

$$\lim_{h \to 0} \frac{f(w+h) - f(w)}{h}, \ h \in \mathbb{C}.$$

❑ Differentiablity $\implies$ continuity.

❑ The elementary functions have the expected derivatives.

❑ Fundamental Theorem of Calculus: Suppose $f : D \to \mathbb{C}$ is a continuous complex function and there exists F such that $F' = f$. Then,

$$\int_\gamma f(z)\,dz = F(\gamma(b)) - F(\gamma(a)).$$

❑ Not all functions have an antiderivative.

❑ A domain D is called path-connected if for each pair $\alpha, \beta \in D$, there exists a contour $\gamma : [a, b] \to D$ such that $\gamma(a) = \alpha$ and $\gamma(b) = \beta$.

❑ (Cauchy-Riemann equations) If $f(z) = u(x, y) + iv(x, y)$ is differentiable, then
$$u_x = v_y \text{ and } v_x = -u_y.$$

❑ The function $\varphi : \mathbb{R}^2 \to \mathbb{R}$ is a harmonic function if it satisfies Laplace's equation:
$$\varphi_{xx} + \varphi_{yy} = 0.$$

❑ For every harmonic u there is a harmonic function v (called a harmonic conjugate of u) such that $f(x + iy) = u(x, y) + i(v, y)$ is complex differentiable.

CHAPTER 9

Differentiation and Integration of Power Series

Power series are central in complex analysis. In Chapter 13 we shall see that, in contrast to real analysis, any complex function differentiable at a point can be given by a power series at that point. Therefore, it is important to study how power series behave under differentiation and integration. In this chapter we show that we can differentiate a power series term-by-term and give a useful condition for integrating term-by-term. Also, we show uniqueness of power series coefficients and their relation to derivatives.

The derivative of a power series is the obvious one

The obvious candidate for the derivative of the series $\sum_{n=0}^{\infty} a_n z^n$ is the one produced by term-by-term differentiation, i.e., $\sum_{n=0}^{\infty} n a_n z^{n-1}$. But, initially, we do not even know that this latter sequence converges. Let us rectify this situation.

Lemma 9.1
The two series $\sum_{n=0}^{\infty} a_n z^n$ and $\sum_{n=0}^{\infty} n a_n z^{n-1}$ have the same radius of convergence.

Proof. First we prove that if $\sum n a_n z^{n-1}$ converges, then $\sum a_n z^n$ converges.
  Suppose that $z \neq 0$ is within the radius of convergence for $\sum n a_n z^{n-1}$. Let

$N \in \mathbb{N}$ be such that $N \geq |z|$ (so $N/|z| \geq 1$). Then,

$$
\begin{aligned}
|na_n z^{n-1}| &= \frac{n|a_n z^n|}{|z|} \\
&\geq |a_n z^n|, \text{ for } n \geq N.
\end{aligned}
$$

By the comparison theorem $\sum_{n=N}^{\infty} a_n z^n$ converges absolutely and therefore so does $\sum_{n=0}^{\infty} a_n z^n$.

Now for the converse. Suppose that R is the radius of convergence for $\sum_{n=0}^{\infty} a_n z^n$. Fix $w \neq 0$ such that $|w| < R$. Choose $r \in \mathbb{R}$ such that $r > 0$ and $|w| + r < R$. Then

$$
\frac{1}{r}(|w| + r)^n = \frac{1}{r} \sum_{k=0}^{n} \binom{n}{k} r^k |w|^{n-k} \geq n|w|^{n-1}.
$$

Thus

$$
|na_n w^{n-1}| \leq \frac{1}{r}|a_n|(|w| + r)^n.
$$

But $|w| + r < R$ and as $\sum a_n z^n$ converges absolutely for all $|z| < R$ and r is fixed we have that $\sum |na_n w^{n-1}|$ converges by the comparison test. Hence, $\sum na_n w^{n-1}$ converges absolutely for all $|w| < R$. $\qquad \square$

The analogue of the statement in Lemma 9.1 is not true in general for series that are not power series.

Example 9.2
The series $S = \displaystyle\sum_{n=1}^{\infty} \frac{\sin nx}{n^2}$ is absolutely convergent for $x \in \mathbb{R}$ but term-by-term differentiation produces $\displaystyle\sum_{n=1}^{\infty} \frac{\cos nx}{n}$. This clearly diverges at $x = 0$.

Now, let's show that $\sum na_n z^{n-1}$ really is the derivative of the series $\sum a_n z^n$.

Theorem 9.3
Suppose that $f(z) = \displaystyle\sum_{n=0}^{\infty} a_n z^n$ has radius of convergence $R > 0$. Then,

$$
f'(z) = \sum_{n=0}^{\infty} na_n z^{n-1}, \text{ for all } |z| < R.
$$

Proof. We know that $g(z) = \sum na_n z^{n-1}$ is well-defined for $|z| < R$. We shall show that $f'(z)$ exists and is equal to $g(z)$.

Suppose that $|z| < R$ and $r \in \mathbb{R}$ is such that $r > 0$ and $|z| + r < R$. Then, for $h \in \mathbb{C}$ with $|h| < r$ we have

$$\left| \frac{(z+h)^n - z^n}{h} - nz^{n-1} \right| = \left| \frac{1}{h} \sum_{k=2}^{n} \binom{n}{k} h^k z^{n-k} \right|$$

$$\leq |h| \sum_{k=2}^{n} \binom{n}{k} |h|^{k-2} |z|^{n-k}, \text{ by triangle inequality,}$$

$$\leq \frac{|h|}{r^2} \sum_{k=2}^{n} \binom{n}{k} r^k |z|^{n-k}, \text{ as } r < |h| \text{ and } k \geq 2,$$

$$\leq \frac{|h|}{r^2} \sum_{k=0}^{n} \binom{n}{k} r^k |z|^{n-k}$$

$$\leq \frac{|h|}{r^2} (|z| + r)^n.$$

So,

$$\left| \frac{f(z+h) - f(z)}{h} - g(z) \right| = \left| \sum_{n=0}^{\infty} a_n \left(\frac{(z+h)^n - z^n}{h} - nz^{n-1} \right) \right|$$

$$\leq \sum_{n=0}^{\infty} |a_n| \left| \frac{(z+h)^n - z^n}{h} - nz^{n-1} \right|$$

$$\leq \sum_{n=0}^{\infty} |a_n| \frac{|h|}{r^2} (|z| + r)^n, \text{ by above,}$$

$$\leq \frac{|h|}{r^2} \sum_{n=0}^{\infty} |a_n| (|z| + r)^n$$

$$= \frac{|h|}{r^2} K, \text{ for some } K \text{ as the series converges,}$$

$$\to 0 \text{ as } h \to 0.$$

Therefore, $f'(z) = g(z) = \sum_{n=0}^{\infty} n a_n z^{n-1}$ as required. $\qquad \square$

The proofs of the lemma and theorem are rather technical but note that there statements are straightforward. They show the following: The derivative of a power series can be found by term-by-term differentation and the resulting power series has the same radius of convergence as the original.

Examples 9.4

(i) The complex functions exp, sin and cos have the expected derivatives. (Exercise.) And since differentiable functions are continuous, these are all continuous functions.

(ii) We know that $\sum_{n=0}^{\infty} z^n = \dfrac{1}{1-z}$ for $|z| < 1$. Differentiating both sides gives

$$\sum_{n=1}^{\infty} n z^{n-1} = \frac{1}{(1-z)^2} \qquad \text{for } |z| < 1.$$

(iii) Let $f(z) = \sum_{n=1}^{\infty} (-1)^{n-1} \dfrac{z^n}{n}$. This converges for $|z| < 1$. We have

$$
\begin{aligned}
f'(z) &= \sum_{n=1}^{\infty} (-1)^{n-1} z^{n-1} \\
&= \sum_{n=1}^{\infty} (-z)^{n-1} \\
&= \sum_{n=0}^{\infty} (-z)^n \\
&= \frac{1}{1-(-z)} \\
&= \frac{1}{1+z}.
\end{aligned}
$$

Thus, if, for clarity, we take $w = z + 1$ we see that

$$\frac{d}{dw} \left(\sum_{n=1}^{\infty} (-1)^{n-1} \frac{(w-1)^n}{n} \right) = \frac{1}{w}.$$

In view of Exercise 8.29(ix) it looks like the series $\sum_{n=1}^{\infty} (-1)^{n-1} \dfrac{(w-1)^n}{n}$ is a good candidate for the Taylor series of the complex logarithm function. That this is the case forms Exercise 13.20(iv).

Infinite differentiability and the coefficients

We have seen that, if f is defined on the disc of convergence, then so is f', and f' is a series. We can differentiate f' term-by-term to get f'' and so on. This suggests the following:

Corollary 9.5

If $f(z) = \sum_{n=0}^{\infty} a_n z^n$ has radius of convergence $R > 0$, then f is infinitely differentiable.

Proof. By induction using the theorem. □

Note that this is to be expected as a similar statement is true for real power series. This result is one of the reasons that power series are so great - we know we can differentiate as many times as we like.

Let us now show that a power series is determined by its derivatives.

Corollary 9.6

If $f(z) = \sum_{n=0}^{\infty} a_n z^n$ has radius of convergence $R > 0$, then

$$a_k = \frac{f^{(k)}(0)}{k!} \text{ for all } k.$$

Proof. By induction we can prove

$$f^{(k)}(z) = \sum_{n=0}^{\infty} n(n-1)\dots(n-k+1)a_n z^{n-k}$$

and this of course holds for $|z| < R$ by Theorem 9.3.

Then,

$$
\begin{aligned}
f^{(k)}(z) &= (k(k-1)\dots(k-k+1)a_k) \\
&\quad + ((k+1)((k+1)-1)\dots((k+1)-k+1)a_{k+1}z) + \dots
\end{aligned}
$$

If we put $z = 0$ into this, then we get $f^{(k)}(0) = k!a_k$. □

Again, you should already know this is true for real series.

Lemma 9.7 (Uniqueness Lemma)

Suppose that, for some $R > 0$,

$$\sum_{n=0}^{\infty} a_n z^n = \sum_{n=0}^{\infty} b_n z^n \text{ for all } |z| < R.$$

Then,

$$a_n = b_n \text{ for all } n.$$

Proof. Let $f(z) = \sum_{n=0}^{\infty} a_n z^n = \sum_{n=0}^{\infty} b_n z^n$. Then, by Corollary 9.6,

$$a_n = \frac{f^{(n)}(0)}{n!} \text{ and } b_n = \frac{f^{(n)}(0)}{n!}.$$

Hence, $a_n = b_n$. □

Note that the Uniqueness Lemma is not trivial. A priori we don't know that we can equate coefficients for infinite series in the same way as for polynomials.

Power series about points other than zero

So far all our power series have been centred at the origin. This is rather limiting and so now we show that we can deal with power series centred at other points.

As an example, let's show how we can differentiate term-by-term.

Proposition 9.8
If $f(z) = \sum a_n(z - z_0)^n$ converges for $|z - z_0| < R$, then $f'(z) = \sum na_n(z - z_0)^{n-1}$ converges for $|z - z_0| < R$.

Proof. Let $h = z - z_0$ and $g(h) = f(z_0 + h)$. Then $f(z) = g(z - z_0)$ and so $g(h) = \sum a_n h^n$ converges for all $|h| = |z - z_0| < R$.

We have

$$f'(z) = \frac{d}{dz}g(z - z_0) = g'(z - z_0)\frac{d}{dz}(z - z_0) = g'(z - z_0) = \sum na_n(z - z_0)^n.$$

$\square$

In general, to translate some result about a power series at 0 to one at the point z_0 we can let h be a new variable defined by $h = z - z_0$. When h is 0, then z is z_0 and vice versa so we really are translating from 0 to z_0. In particular, if $|h| < R$ for some R, then we are defining a disc around z_0.

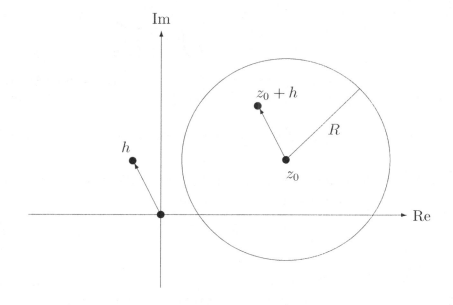

Suppose that for a function f we have $f(z_0 + h) = \sum_{n=0}^{\infty} a_n h^n$, with the series convergent for $|h| < R$. Then, $f(z) = \sum_{n=0}^{\infty} a_n(z - z_0)^n$, which converges for $|h| = |z - z_0| < R$. Thus, if we want to say something about $f(z)$ at z_0 we can use the power series in terms of h as in the previous proposition.

Convergence on the radius

We have seen that a power series and its derivative have the same radius of convergence. This statement is about convergence *within* a circle. An obvious question is what happens on the circle, i.e., when $|z| = R$? Unfortunately, in this case we can't say anything very general.

Example 9.9
The power series $\sum_{n=1}^{\infty} \dfrac{z^n}{n^2}$ converges for $|z| < 1$ and diverges for $|z| > 1$ by the ratio test. For $|z| = 1$ we have

$$\sum_{n=1}^{\infty} \left| \frac{z^n}{n^2} \right| = \sum_{n=1}^{\infty} \frac{|z|^n}{n^2} = \sum_{n=1}^{\infty} \frac{1^n}{n^2} = \sum_{n=1}^{\infty} \frac{1}{n^2}.$$

Thus on the circle $|z| = 1$ the series converges. Hence, in summary, $\sum_{n=1}^{\infty} \dfrac{z^n}{n^2}$ converges for $|z| \leq 1$.

The term-by-term differentiation of this series gives

$$\frac{d}{dz} \left(\sum_{n=1}^{\infty} \frac{z^n}{n^2} \right) = \sum_{n=1}^{\infty} \frac{d}{dz} \left(\frac{z^n}{n^2} \right) = \sum_{n=1}^{\infty} \frac{z^{n-1}}{n}.$$

From Theorem 9.3 we know that this converges for $|z| < 1$. Does this derivative converge on the circle $|z| = 1$ like the original series? The answer is no: Let $z = 1$, then

$$\sum_{n=1}^{\infty} \frac{z^{n-1}}{n} = \sum_{n=1}^{\infty} \frac{1}{n}.$$

We know that this does not converge. On the other hand the derivative does converge for $z = -1$ (by the Alternating Series test):

$$\sum_{n=1}^{\infty} \frac{z^{n-1}}{n} = \sum_{n=1}^{\infty} \frac{(-1)^{n-1}}{n}.$$

Thus on the circle $|z| = 1$ the derivative converges for some points and diverges for others even though the original series converged on the circle.

Integration of power series

The following theorem gives us conditions to ensure that we can integrate a series term-by-term. The statement is often known as the **Weierstrass M-test**. (This test holds in more generality and is used to show what is known as uniform convergence of a series of functions.)

Theorem 9.10 (Term-by-term integration of series)
Let γ be a contour in a domain D. Let $f : D \to \mathbb{C}$ and $f_k : D \to \mathbb{C}$ be continuous complex functions, $k \in \mathbb{N}$. Suppose that

(i) $\displaystyle\sum_{k=0}^{\infty} f_k(z)$ converges to $f(z)$ for all $z \in \gamma^*$;

(ii) M_k are real constants such that $|f_k(z)| \leq M_k$ for all $z \in \gamma^*$;

(iii) $\displaystyle\sum_{k=0}^{\infty} M_k$ converges.

Then,

$$\sum_{k=0}^{\infty} \int_{\gamma} f_k(z)\,dz = \int_{\gamma} \sum_{k=0}^{\infty} f_k(z)\,dz = \int_{\gamma} f(z)\,dz.$$

Proof. The second equality is obvious from (i). We show the first. Let $M =$

$\sum_{k=0}^{\infty} M_k$. We have

$$
\begin{aligned}
\left| \int_{\gamma} f(z)\, dz - \sum_{k=0}^{n} \int_{\gamma} f_k(z)\, dz \right|
&= \left| \int_{\gamma} \left(f(z) - \sum_{k=0}^{n} f_k(z) \right) dz \right| \\
&\leq \sup_{z \in \gamma^*} \left\{ \left| f(z) - \sum_{k=0}^{n} f_k(z) \right| \right\} \times L(\gamma) \\
&= \sup_{z \in \gamma^*} \left\{ \left| \sum_{k=n+1}^{\infty} f_k(z) \right| \right\} \times L(\gamma) \\
&\leq \sup_{z \in \gamma^*} \left\{ \sum_{k=n+1}^{\infty} |f_k(z)| \right\} \times L(\gamma) \\
&\leq \left(\sum_{k=n+1}^{\infty} M_k \right) \times L(\gamma) \\
&= \left(M - \sum_{k=0}^{n} M_k \right) \times L(\gamma) \\
&\to 0 \times L(\gamma), \ \text{as } n \to \infty, \\
&= 0.
\end{aligned}
$$

Thus, the first equality holds. $\qquad\qquad\square$

Example 9.11

For any contour γ the integral $\int_{\gamma} e^z\, dz$ can be calculated by term-by-term integration of the series for e^z.

Let $f_k(z) = \dfrac{z^k}{k!}$, then $e^z = \sum_{k=0}^{\infty} f_k(z)$, so condition (i) is fulfilled. We have

$$
|f_k(z)| = \left| \frac{z^k}{k!} \right| = \frac{|z|^k}{k!}.
$$

But $|z| = |\gamma(t)|$ and γ is a continuous map from $[a, b]$ to $\mathbb{C}$ so its image must lie within an origin-centred circle of radius R, for some large enough R. Thus $|z| \leq R$ for all $z \in \gamma^*$. (Alternatively, we can argue that $|\gamma|$ is a continuous map from $[a, b]$ to $\mathbb{R}$ and from the theory of real functions it is bounded, i.e., $|\gamma(t)| < R$.)

Hence, let $M_k = \dfrac{R^k}{k!}$, then $|f_k(z)| \leq M_k$ for all $z \in \gamma^*$. Therefore, condition (ii) holds.

Also,

$$\sum_{k=0}^{\infty} M_k = \sum_{k=0}^{\infty} \frac{R^k}{k!} = e^R,$$

so $\sum_{k=0}^{\infty} M_k$ converges. Condition (iii) holds.

Thus,

$$\int_{\gamma} e^z \, dz = \int_{\gamma} \sum_{k=0}^{\infty} \frac{z^k}{k!} = \sum_{k=0}^{\infty} \int_{\gamma} \frac{z^k}{k!} \, dz = \sum_{k=0}^{\infty} \frac{z^{k+1}}{(k+1)!} \Big|_{\gamma(a)}^{\gamma(b)} = e^z \Big|_{\gamma(a)}^{\gamma(b)}.$$

Exercises

Exercises 9.12

(i) Prove using term-by-term differentiation that exp, sin and cos have the expected derivatives. That is,

$$\frac{d}{dz} e^z = e^z, \qquad \frac{d}{dz} \sin z = \cos z \qquad \frac{d}{dz} \cos z = -\sin z.$$

(ii) Provide the details for the proof of Corollary 9.5.

(iii) Show that the derivative of $f(z) = \sum_{n=0}^{\infty} (-1)^n \frac{z^{2n+1}}{2n+1}$, $|z| < 1$, is $\frac{1}{1+z^2}$.

(Over the reals this is the series for arctan.)

Generalize this result so that the derivative is $\frac{1}{1+z^k}$ for k a positive integer.

(iv) Give a power series for $\frac{1}{(1-z)^k}$ where k is a natural number.

(v) Suppose that $f(z) = \sum_{n=0}^{\infty} a_n (z - z_0)^n$ converges for $|z - z_0| < R$ for some $R > 0$ and γ is a contour such that $\gamma^* \subset \{z : |z - z_0| < R\}$. Show that

$$\int_{\gamma} f(z) \, dz = \sum_{n=0}^{\infty} \int_{\gamma} a_n (z - z_0)^n \, dz.$$

(vi) Calculate B_n for $n = 0, \ldots, 4$ where

$$\frac{z}{e^z - 1} = \sum_{n=0}^{\infty} \frac{B_n}{n!} z^n.$$

The number B_n are called **Bernoulli numbers**

Summary

Let $f(z) = \sum_{n=0}^{\infty} a_n z^n$ have radius of convergence $R > 0$.

❑ f can be differentiated term-by-term to get the derivative.

❑ f' has the same radius of convergence as f.

❑ f is infinitely differentiable.

❑ The coefficients a_n are unique and $a_n = \dfrac{f^{(n)}(0)}{n!}$ for all n.

❑ We can translate these results to series of the following form:

$$f(z) = \sum_{n=0}^{\infty} a_n (z - z_0)^n,$$

for $|z - z_0| < R$.

❑ We can integrate term-by-term complex series that satisfy a Weierstrass M-test type condition.

CHAPTER 10

Winding Numbers

In the next chapter we state one of the most remarkable theorems in mathematics. To do this we need the notions of winding number and interior point. Both these notions are intuitively simple though the definition of winding number at first appears unintuitive.

Definition 10.1
Let γ be a closed contour and w a point not on γ^*. Then the **winding number** of γ about w, denoted $n(\gamma, w)$, is

$$n(\gamma, w) = \frac{1}{2\pi i} \int_\gamma \frac{dz}{z - w}.$$

This concept is made more intuitive by the following lemma.

Lemma 10.2 (Winding Number Lemma)
Let γ be a closed contour, $w \in \mathbb{C}\backslash\gamma^*$. Then the winding number of γ about w is an integer and represents the net number of times that γ winds about w, with anticlockwise counted positively.

First some examples to make this clearer.

Example 10.3
The winding numbers for points in the regions enclosed by the contour are shown below.

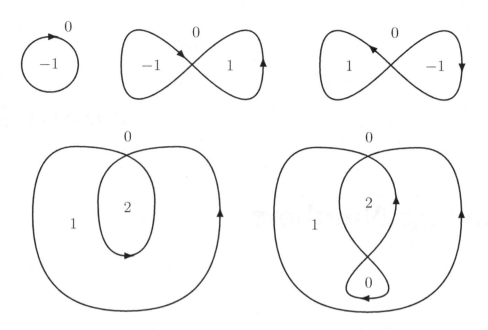

Exercise 10.4
Find the winding numbers for points in the various following regions.

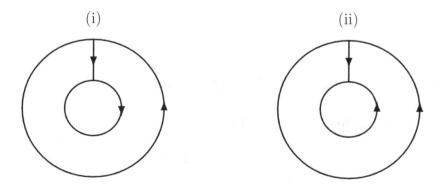

(i)   (ii)

Proof (of Winding Lemma). If we write $\gamma(t) = w + r(t)e^{i\theta(t)}$, then $r(t)$ is a continuous, piecewise continuous differentiable functions on $[a, b]$ and $\theta(t)$ can be chosen to be so. We have,

$$\int_\gamma \frac{dz}{z-w} = \int_a^b \frac{\gamma'(t)}{\gamma(t)-w}\, dt$$

$$= \int_a^b \frac{r'(t)e^{i\theta(t)} + i\theta'(t)r(t)e^{i\theta(t)}}{r(t)e^{i\theta(t)}}\, dt$$

$$= \int_a^b \frac{r'(t)}{r(t)}\, dt + i \int_a^b \theta'(t)\, dt$$

$$= [\log r(t)]_a^b + i\, [\theta(t)]_a^b$$

$$= \log r(b) - \log r(a) + i\,(\theta(b) - \theta(a)).$$

Since γ is a closed contour we have $r(a)e^{i\theta(a)} = r(b)e^{i\theta(b)}$. Equating the moduli and arguments we get $r(a) = r(b)$ and $\theta(b) = \theta(a) + 2\pi n$ for some $n \in \mathbb{Z}$.

Putting these results into the integral above gives

$$\int_\gamma \frac{dz}{z-w} = 2\pi i n$$

and hence the winding number is an integer.

As t goes from a to b, the net increase in $\arg(\gamma(t) - w)$ measures 2π times the number of times that γ winds round w. But,

$$\frac{\arg(\gamma(t) - w)|_{t=a}^{t=b}}{2\pi} = \frac{\theta(b) - \theta(a)}{2\pi} = n = n(\gamma, w).$$

$\square$

Theorem 10.5

(i) *Suppose that γ is a closed contour and $w \in \mathbb{C}\backslash\gamma^*$. Then,*

$$n(-\gamma, w) = -n(\gamma, w).$$

(ii) *Suppose that γ_1 and γ_2 are closed contours, so that their join can be taken. Then, for all $w \in \mathbb{C}\backslash(\gamma_1 + \gamma_2)^*$,*

$$n(\gamma_1 + \gamma_2, w) = n(\gamma_1, w) + n(\gamma_2, w).$$

Proof. Exercises. $\square$

Figure 10.1: Calculate the winding number by eye.

An easy method of calculation

It is easy to calculate winding numbers 'by eye', rather than calculation of a contour integral.

Consider the contour drawn in Figure 10.1 that is traversed once in the direction indicated. A straight line has been drawn from one side to the other that does not pass through any self-intersections of the curve. We can trace round the curve noting the direction of travel as we cross the straight line.

Start at the far right side of the straight line. Obviously the winding number of a point there is 0. As we go from right to left on the line we will cross the contour. We apply the following rules:

❏ If we cross the contour so that it is travelling *up*, then the winding number *increases* by 1.

❏ If we cross the contour so that it is travelling *down*, then the winding number *decreases* by 1.

In the above diagram, the contour is travelling up when we first meet it, so the point in the region we pass into have winding number 1. At the next crossing the contour is going down, and so the winding number of points in the next region is 0. We can carry this out for all regions on the line. To get other regions we can use different lines or go from one region to the other applying the rules.

Exercise 10.6
Find all the winding numbers for the points in Figure 10.1. A useful starting line has been drawn in already.

Justification of the method

Assume that the contour is traversed once and consider two points w and z that lie on opposite sides of a contour line that goes up. We can assume that the contour starts and finishes at a point near w and z. (If it didn't, then we can do some work with reparametrisations and joins.)

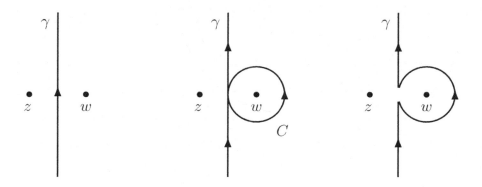

Now, take a loop C round w that starts and ends at the start of γ. Then, $n(\gamma + C, w) = n(\gamma, z)$, as the diagram shows: These winding numbers are the same as the middle and right pictures are just minor perturbations of each other. Thus,

$$
\begin{aligned}
n(\gamma, z) &= n(\gamma + C, w) \\
&= n(\gamma, w) + n(C, w), \text{ by Theorem 10.5}, \\
&= n(\gamma, w) + 1.
\end{aligned}
$$

This shows that if we pass from w to z, then the winding number increases by 1. Similarly, one can show a decrease by 1 for a contour locally heading down.

The case where the contour, or part of the contour, is traversed more than once is straightforward. Suppose that at the crossing point the contour is traversed a net k times going up, then the winding number goes up by k. Here net means that, for example, if the curves goes up 5 times and down 2 times on that portion, then the curve goes up by 3 in total.

For example, consider the contour from Exercise 5.18(iv). This is pictured below. The inner circle is traversed twice so instead of increasing or decreasing the winding number by 1 or -1 it changes by 2 or -2. The winding numbers for each region are indicated.

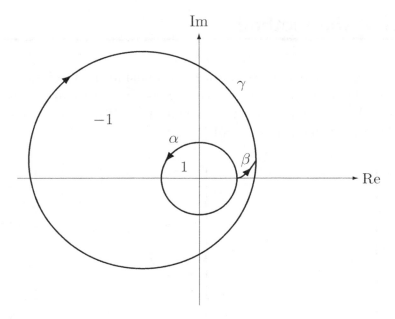

Interior points

It is intuitively clear that a closed curve has an interior and exterior. However, mathematically stating this and proving it is not trivial. We shall avoid this problem by defining the exterior points of the image of a contour to be those that have winding number zero and interior points to be those with non-zero winding numbers.

First let us show that there exists an (unbounded) open set whose points have winding number zero. These we think of as being exterior points.

Theorem 10.7
Let γ be a closed contour. There exists an open set D such that $n(\gamma, w) = 0$ for

all $w \in D$.

Proof. Let R be a bound on $|\gamma(t)|$. This exists as the function $|\gamma(t)|$ is a continuous real valued function on a closed interval and hence attains its bounds. Define D to be the set of all points $w \in \mathbb{C}$ such that $|w| > R + L(\gamma)$. This set is open. Furthermore, $n(\gamma, w) = 0$ for all $w \in D$ as we now show. For all $z \in \gamma^*$ we have

$$|w - z| \geq |w| - |z| \geq R + L(\gamma) - R = L(\gamma).$$

So

$$\left| \frac{1}{z - w} \right| = \left| \frac{1}{w - z} \right| \leq \left| \frac{1}{L(\gamma)} \right| = \frac{1}{L(\gamma)}.$$

Hence,

$$
\begin{aligned}
|n(\gamma, w)| &= \left| \frac{1}{2\pi i} \int_\gamma \frac{1}{z - w} \, dz \right| \\
&\leq \frac{1}{2\pi} \left| \int_\gamma \frac{1}{z - w} \, dz \right| \\
&\leq \frac{1}{2\pi} \cdot \frac{1}{L(\gamma)} L(\gamma), \text{ by the Estimation Lemma,} \\
&< 1.
\end{aligned}
$$

So $n(\gamma, w) = 0$.

That completes the proof but it should be remarked that we can do better than this. Define $\widetilde{D}$ to be the set of all points $\widetilde{w} \in \mathbb{C}$ such that there exists a contour Γ from w to $\widetilde{w}$ such that Γ^* and γ^* do not intersect. Then $n(\gamma, \widetilde{w}) = n(\gamma, w) = 0$ by Exercise 10.11(i). The set $\widetilde{D}$ is open but the proof of this is non-trivial. $\square$

From the proof we can see that this open set consists of the points that one would think of as exterior to the contour. We shall mostly be interested in points with non-zero winding number and so make the following definition.

Definition 10.8

Let γ be a closed contour. The point $w \in \mathbb{C} \backslash \gamma^*$ is an **interior point** of γ if $n(\gamma, w) \neq 0$. We denote the set of interior points of γ by $\mathrm{Int}(\gamma)$.

Example 10.9

The interior points of the contour in the next diagram are shaded and the windings numbers are given.

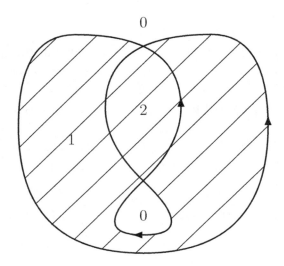

Example 10.10
Let $\gamma(t) = w + Re^{2\pi it}$, $0 \le t \le 1$. Then, $\mathrm{Int}(\gamma) = \{z \,:\, |z - w| < R\}$.

Exercises

Exercises 10.11

(i) Let γ be a closed curve. Suppose that w_1 and w_2 are points such that a contour α starts at w_1 and ends at w_2 without crossing the image of γ. Show, using the definition of winding number, that $n(\gamma, w_1) = n(\gamma, w_2)$.

(ii) Prove the statements in Theorem 10.5.

(iii) Find the winding numbers for points in the all the regions.

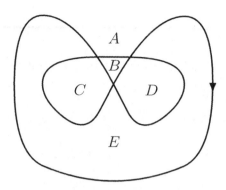

(iv) Prove that the set of interior points of a contour is an open set.

(v) Sketch the contour $\gamma(t) = 5e^{it} \cos \dfrac{2t}{3}$, for $0 \le t \le 12\pi$. Calculate $n(\gamma, 1)$, $n(\gamma, -1)$, $n(\gamma, 2 + 2i)$ and $n(\gamma, 4)$.

(vi) Let γ be the standard contour giving the circle or radius a. Let $b > a$.

   (a) Calculate the integral $\displaystyle\int_\gamma \frac{dz}{z - b}$.

   (b) Deduce that
$$\int_0^{2\pi} \frac{ab \sin t}{a^2 - 2ab \cos t + b^2}\, dt = 0$$
   and
$$\int_0^{2\pi} \frac{a^2 - ab \cos t}{a^2 - 2ab \cos t + b^2}\, dt = 0.$$

(vii) Let $\Psi(f) = \dfrac{f'}{f}$ where f is a differentiable function on the open set D.

   (a) Prove that $\Psi(fg) = \Psi(f) + \Psi(g)$ for all differentiable functions f and g.

   (b) Deduce that if $f(z) = \prod_{j=1}^m (z - a_j)$ where $a_j \in \mathbb{C}$, then for any contour γ with $a_j \notin \gamma^*$ for all j, then
$$\frac{1}{2\pi i} \int_\gamma \frac{f'(z)}{f(z)}\, dz = \sum_{j=1}^m n(\gamma, a_j).$$

Summary

❏ The winding number is the number of times a contour wraps round a point in an anticlockwise direction.

❏ Let γ be a closed contour, $w \in \mathbb{C} \backslash \gamma^*$. Then

$$n(\gamma, w) = \frac{1}{2\pi i} \int_\gamma \frac{dz}{z - w}.$$

❏ It is easy to calculate a winding number by eye.

❏ An interior point is any point with non-zero winding number.

CHAPTER 11

Cauchy's Theorem

We now come to the fundamental theorem in complex analysis. There is no analogue in real analysis and it has far reaching, deep and surprising consequences, including applications to real variable problems. The rest of the book relies on this theorem.

A proof of the theorem is usually the most difficult part of an introductory complex analysis course. The most general proofs are often complicated, long, and do not give any insight into why the result is true. In this chapter we will state the theorem in generality and give a reasonably simple proof for a case that will be sufficient for all the later results in the book. A full proof is given in Appendix A.

Theorem 11.1 (Cauchy's Theorem)
Let $D \subseteq \mathbb{C}$ be an open set, and $f : D \to \mathbb{C}$ be a differentiable complex function. Let γ be a closed contour such that γ and its interior points are in D.

Then, $\displaystyle\int_\gamma f = 0.$

Remarks 11.2
  (i) This is truly a great theorem. It refers to *any* open set in $\mathbb{C}$, *any* differentiable function on D, and *any* contour with all interior points in D. And it says that *any* integral arising from this is zero. Thus, weak assumptions lead to a strong conclusion.

  (ii) It is necessary that the interior of γ lies within D. Let $D = \mathbb{C}\backslash\{w\}$ for any point in $w \in \mathbb{C}$. Suppose that $f(z) = 1/(z - w)$ and that γ is *any* contour that has w in its interior. Then, f is differentiable on D and

$$\int_\gamma f = \int_\gamma \frac{dz}{z-w} = 2\pi i\, n(\gamma, w) \neq 0.$$

(iii) Note that we can't just use the Fundamental Theorem of Calculus as in Corollary 8.11 since we don't know whether f has an antiderivative on D. (And usually it won't have.)

(iv) At first sight it may appear that the theorem will only tell us about the behaviour of differentiable functions. However, it has strong implications for non-differentiable functions as well as we will see later in Cauchy's Residue Theorem.

Exercise 11.3
Using the contours in Exercises 2 Question 1, to which of the following integrals does Cauchy's theorem apply? (There is no need to evaluate them.)

$$\int_{\gamma_1} |z|^2\, dz, \qquad \int_{\gamma_1} \frac{z^2}{z-2}\, dz, \qquad \int_{\gamma_2} z\, dz, \qquad \int_{\gamma_2+\gamma_3} \sin\left(\frac{1}{z-1}\right) dz.$$

Proof of a simple version of Cauchy's Theorem

We shall now begin to prove a weaker version of the theorem.

Definition 11.4
A **square** is a set of the form

$$\{z \in \mathbb{C} \mid a \le \mathrm{Re}(z) \le b, c \le \mathrm{Im}(z) \le d$$
$$\text{for some } a, b, c, d \in \mathbb{R} \text{ with } b - a = d - c > 0\}.$$

Lemma 11.5 (Nested Squares Lemma)
Suppose that $Q_1 \supset Q_2 \supset Q_3 \supset \ldots$ is a sequence of squares, then $\bigcap\limits_{n=1}^{\infty} Q_n \neq \emptyset$.

Proof. Let $Q_n = \{z \in \mathbb{C} \mid a_n \le \mathrm{Re}(z) \le b_n, c_n \le \mathrm{Im}(z) \le d_n\}$. By assumption $[a_1, b_1] \supset [a_2, b_2] \supset [a_3, b_3] \supset \ldots$ and so we have

$$a_1 \le a_2 \le a_3 \le \cdots \le a_n < \cdots < b_n \le b_{n-1} \le \cdots \le b_2 \le b_1$$

for all n. Thus (a_n) is a bounded increasing sequence and hence converges to some a_0 with $a_n \le a_0$. Similarly the sequence b_n has a limit b_0 with $b_0 \le b_n$. As

$a_n < b_n$ we have $a_0 \le b_0$. Now $a_0 \in [a_n, b_n]$ for all n as $a_n \le a_0 \le b_0 \le b_n$. So $a_0 \in \bigcap_{n=1}^{\infty} [a_n, b_n]$. Similarly there exists $c_0 \in \bigcap_{n=1}^{\infty} [c_n, d_n]$ for some c_n, d_n.

Therefore $a_0 + ic_0 \in \bigcap_{i=1}^{\infty} Q_n$.   □

Lemma 11.6
Let γ be a simple closed contour made of a finite number of lines and arcs in the open set D with $\widetilde{D} = \gamma^* \cup \mathrm{Int}(\gamma) \subseteq D$. Let Q be a square in $\mathbb{C}$ bounding $\widetilde{D}$ and $f : D \to \mathbb{C}$ be analytic. Then for any $\epsilon > 0$ there exists a subdivision of Q into a grid of squares so that for each square Q_j in the grid with $Q_j \cap \widetilde{D} \ne \emptyset$ there exists a $z_j \in Q_j \cap \widetilde{D}$ such that

$$\left| \frac{f(z) - f(z_j)}{z - z_j} - f'(z_j) \right| < \epsilon \quad \text{for all } z \in Q_j \cap \widetilde{D}.$$

Proof. The set up looks like the following diagram.

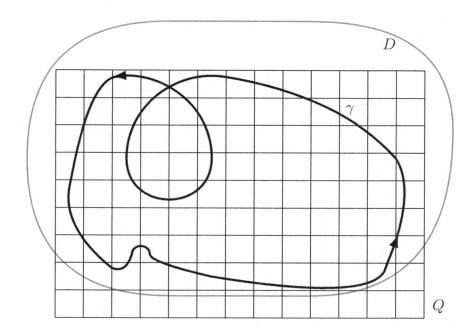

For a contradiction assume that the statement is not true. Let $Q_1 = Q$ and divide Q_1 into 4 equal-sized squares. At least one of these squares will not satisfy the required condition in the lemma. Let Q_2 be such a square. Repeat the process to produce an infinite sequence of squares with $Q_1 \supset Q_2 \supset Q_3 \supset \ldots$. By the Nested Squares Lemma there exists $z_j \in \bigcap_{n=1}^{\infty} Q_n$.

As f is differentiable there exists $\delta > 0$ such that

$$\left| \frac{f(z) - f(z_j)}{z - z_j} - f'(z_j) \right| < \epsilon$$

for $|z - z_j| < \delta$. But as the size of the squares becomes arbitrarily small there must exist Q_N such that Q_N is contained in the disk $|z - z_j| < \delta$. This is a contradiction. $\qquad\square$

Remark 11.7

From this we can deduce that we can subdivide the square Q by squares with side length $S/2^N$ where S is the length of the side of Q and N is a natural number.

Theorem 11.8 (A Simple Version of Cauchy's Theorem)

Let $D \subseteq \mathbb{C}$ be a open set and $f : D \to \mathbb{C}$ be a differentiable function. Let γ be a simple closed contour made of a finite number of lines and arcs such that $\gamma^* \cup Int(\gamma) \subset D$. Then

$$\int_{\gamma} f(z) dz = 0.$$

Proof. Given $\epsilon > 0$ there exists a grid of squares covering $\gamma^* \cup Int(\gamma)$ as in Lemma 11.6. Let $\{S_j\}_{j=1}^{n}$ be the set of squares such that $S_j \cap (\gamma^* \cup Int(\gamma)) \neq \emptyset$ and let $\{z_j\}_{j=1}^{n}$ be the set of distinguished points in the lemma.

Define $g_j : D \to \mathbb{C}$ by

$$g_j(z) = \begin{cases} \dfrac{f(z) - f(z_j)}{z - z_j} - f'(z_j), & z \neq z_j \\ 0, & z = z_j \end{cases}$$

Then as f is differentiable, g_j is continuous (and hence integrable).

Without loss of generality we can assume that γ is positively oriented. Let C_j be the union of positively oriented contours giving the boundary of $S_j \cap (\gamma^* \cup Int(\gamma))$. Since γ is made of a finite number of lines and arcs C_j will itself be the union of a finite number of lines and arcs. For S_j such that $S_j \cap \gamma^* = \emptyset$, C_j^* is just the boundary of a square. An example of a C_j consisting of three disjoint contours can be seen in the following diagram.

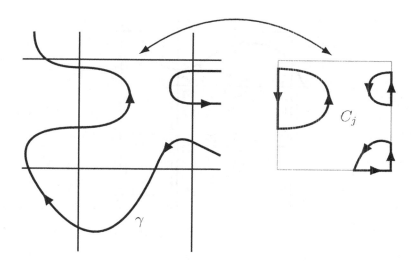

On S_j we have

$$f(z) = f(z_j) + (z - z_j)f'(z_j) + (z - z_j)g_j(z). \tag{11.1}$$

As $f(z_j) + (z - z_j)f'(z_j)$ is the derivative of

$$(z - z_j)f(z_j) + \frac{(z - z_j)^2}{2}f'(z_j)$$

by the Fundamental Theorem of Calculus and the fact that C_j is closed we get

$$\int_{C_j} f(z_j) + (z - z_j)f'(z_j)dz = 0. \tag{11.2}$$

Now,

$$\int_\gamma f(z)dz = \sum_{j=1}^n \int_{C_j} f(z)dz$$

and edges of touching squares will cancel. This is pictured in the following diagram.

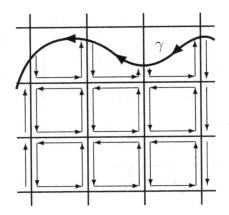

So

$$\left| \int_\gamma f(z)dz \right| = \left| \sum_{j=1}^{n} \int_{C_j} f(z)dz \right|$$

$$\leq \sum_{j=1}^{n} \left| \int_{C_j} f(z)dz \right|$$

$$= \sum_{j=1}^{n} \left| \int_{C_j} (z - z_j)g_j(z)dz \right| \qquad \text{by (11.1) and (11.2)}.$$

We now estimate each of the integrals in the sum.

Let s be the length of the side of the squares. For $z, z_j \in S_j$ we have

$$|(z - z_j)g_j(z)| < \sqrt{2}s\epsilon$$

because $|z - z_j| \leq \sqrt{2}s$ as S_j is a square and $|g_j(z)| < \epsilon$ as the grid of squares satisfies the conclusion of the lemma.

Let l_j be the length of the curve(s) in $S_j \cap \gamma^*$ (the length may be zero). Then

$$L(C_j) \leq l_j + 4s.$$

Hence, by the Estimation Lemma

$$\left| \int_{C_j} (z - z_j)g_j(z)dz \right| < \sqrt{2}s\epsilon(l_j + 4s).$$

Therefore

$$\left| \int_\gamma f(z)dz \right| < \sum_{j=1}^{n} \sqrt{2} s \epsilon (l_j + 4s)$$

$$= \sqrt{2} \epsilon \sum_{j=1}^{n} (sl_j + 4s^2)$$

$$= \sqrt{2} \epsilon (sL(\gamma) + 4A)$$

where A is the area of all the squares $\{S_j\}_{j=1}^{n}$. Now s is less than or equal to the length S of the side of the original square enclosing D. Hence,

$$\left| \int_\gamma f(z)dz \right| < \sqrt{2} \epsilon (SL(\gamma) + 4S^2).$$

As ϵ was arbitrary and S and $L(\gamma)$ are fixed we have

$$\int_\gamma f(z)dz = 0.$$

$\square$

Remark 11.9
The reason for having a finite number of lines and arcs is because then we know that each C_j^* consists of a finite number of disjoint contours. If we allowed any closed contour, then in principle it is possible to get an infinite number of pieces in a C_j^*. Hence we would have to worry about an infinite sum. This problem is not impossible to overcome but at the moment we wish to have a version of the theorem good enough for our proofs and applications.

Example 11.10
Consider the following diagram of contours.

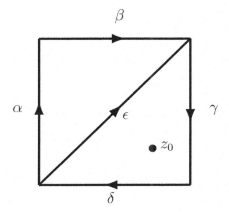

Let f be a function analytic on and within the square except at the point z_0. We can show that

$$\int_{\alpha+\beta+\gamma+\delta} f \, dz = \int_{\gamma+\delta+\epsilon} f \, dz$$

as follows.

As f is complex differentiable $\int_{\alpha+\beta-\epsilon} f = 0$ and so

$$\int_{\alpha+\beta+\gamma+\delta} f = \int_{\alpha+\beta-\epsilon+\epsilon+\gamma+\delta} f = \int_{\alpha+\beta-\epsilon} f + \int_{\epsilon+\gamma+\delta} f = \int_{\gamma+\delta+\epsilon} f.$$

In theory the integrals of a function over two separate contours could be distinct. The following proposition shows that, in certain circumstances, the integrals are independent of the path taken.

Proposition 11.11

Suppose that α and β are contours such that their start points coincide and their end points coincide. Further suppose that $f : D \to \mathbb{C}$ is a function analytic on an open set D such that $\alpha - \beta$ and the interior of $\alpha - \beta$ are in D. Then,

$$\int_\alpha f(z) \, dz = \int_\beta f(z) \, dz.$$

Proof. The result follows from

$$\int_\alpha f(z) \, dz - \int_\beta f(z) \, dz = \int_{\alpha-\beta} f(z) \, dz = 0.$$

$\square$

Exercises

Exercises 11.12

(i) Let

$$D_1 = \mathbb{C},$$
$$D_2 = \{z \in \mathbb{C} \mid |z| \le 4\},$$
$$D_3 = \{z \in \mathbb{C} \mid |z| < 4\},$$
$$D_4 = \{z \in \mathbb{C} \mid \mathrm{Im}(z) \ge -1\},$$
$$D_5 = \{z \in \mathbb{C} \mid \mathrm{Im}(z) \ge -10\},$$
$$f_1(z) = z^2 + 3z - 5,$$
$$f_2(z) = \frac{1}{z^2 + 4},$$
$$f_3(z) = \frac{1}{z + 2i},$$
$$C_r(t) = re^{it}, 0 \le t \le 2\pi,$$
$$\Gamma = \text{boundary of } \{z \in \mathbb{C} \mid |z| \le 10, \ \mathrm{Im}(z) \ge 0\}.$$
$$\beta(t) = t, -1 \le t \le 1.$$

Consider the following open sets, functions and contours. Which of the combination satisfy the assumptions of Cauchy's Theorem? Where the assumptions are not satisfied, then explicitly explain why.

(a) Let $D = D_2$, $f = f_1$, $\gamma = C_1$.

(b) Let $D = D_3$, $f = f_1$, $\gamma = C_1$.

(c) Let $D = D_3$, $f = f_1$, $\gamma = C_4$.

(d) Let $D = D_3 \setminus \{-2i\}$, $f = f_2$, $\gamma = C_1$.

(e) Let $D = D_3 \setminus \{-2i\}$, $f = f_3$, $\gamma = C_1$.

(f) Let $D = D_3 \setminus \{-2i\}$, $f = f_3$, $\gamma = C_3$.

(g) Let $D = D_1$, $f = f_1$, $\gamma = \beta$.

(h) Let $D = D_1$, $f = f_1$, $\gamma = \beta + C_1 - \beta$.

(i) Let $D = D_5$, $f = f_2$, $\gamma = \Gamma$.

(j) Let $D = D_4 \setminus \{2i\}$, $f = f_2$, $\gamma = \Gamma$.

(k) Let $D = D_4$, $f = f_3$, $\gamma = \Gamma$.

(l) Let $D = D_5$, $f = f_3$, $\gamma = \Gamma$.

(ii) (Existence of antiderivatives II.) Let U be an open disc in $\mathbb{C}$ centred at w and $f : U \to \mathbb{C}$ be a differentiable function on U. Let $\Gamma(z)$ be the straight line contour from w to z for each $z \in U$. That is, $\Gamma(z)(t) = (t-1)w + tz$ for all $0 \le t \le 1$.

Show that $F(z) = \int_{\Gamma(z)} f(\zeta) \, d\zeta$ is an antiderivative of f on U, i.e., $F'(z) = f(z)$ for all $z \in U$.

Why is this not a counterexample to Remark 11.2(iii)?

(iii) Let D be a path connected open set in $\mathbb{C}$ and $f : D \to \mathbb{C}$ be a differentiable function. Let γ_1 and γ_2 be closed contours in D such that $n(\gamma_1, z) = -n(\gamma_2, z)$ for all $z \notin D$. Show that

$$\int_{\gamma_1} f(z) \, dz = -\int_{\gamma_2} f(z) \, dz.$$

[Hint: Consider a contour from a point of image of γ_1 to one of γ_2.]

(iv) Suppose that $f : D \to \mathbb{C}$ is a complex analytic function on the open set D such that f is complex differentiable except at $w \in D$. Prove there exists $R > 0$ such that for all $0 < r < R$ we have

$$\int_{C_R} f(z) \, dz = \int_{C_r} f(z) \, dz.$$

(This is an important exercise and a proof will be given later. Nonetheless, is worthwhile to work this out.)

(v) Suppose that γ_1 and γ_2 are two contours such that $\gamma_2^* \subset \mathrm{Int}(\gamma_1)$. Let f be a function that is differentiable on γ_1^* and γ_2^* and $\mathrm{Int}(\gamma_1) \backslash \mathrm{Int}(\gamma_2)$.

Prove that, if $n(\gamma_1, w) = n(\gamma_2, w)$ for all $w \in \mathrm{Int}(\gamma_2)$, then

$$\int_{\gamma_1} f(z) \, dz = \int_{\gamma_2} f(z) \, dz.$$

[Warning: Don't forget to carefully construct an open set upon which f is differentiable.]

(vi) Let $\gamma = \gamma_1 + \gamma_2 + \gamma_3$ where γ_1 is the contour given by going 0 to R on the real line, γ_2 goes anticlockwise in a circle arc to $e^{i\pi/4}R$ and γ_3 goes in a straight line from $e^{i\pi/4}R$ back to the origin. (This is called a **slice of pie contour** or **sector contour** due to its shape.)

(a) Show that $\int_\gamma e^{-z^2} \, dz = 0$.

(b) Show that $\int_{\gamma_2} e^{-z^2} \, dz \to 0$ as $R \to \infty$.

(c) Show that

$$\int_{\gamma_3} e^{-z^2} \, dz = -e^{i\pi/4} \int_0^R e^{-ix^2} \, dx.$$

(d) Hence, using the well-known equality $\int_{-\infty}^{\infty} e^{-x^2} \, dx = \sqrt{\pi}$, deduce the **Fresnel integrals**:

$$\int_0^{\infty} \sin(x^2) \, dx = \sqrt{\frac{\pi}{8}},$$
$$\int_0^{\infty} \cos(x^2) \, dx = \sqrt{\frac{\pi}{8}}.$$

(vii) If we add the assumption that f' is continuous to the statement of Cauchy's Theorem then the resulting theorem can be proved quite quickly using Green's Theorem. Either prove this or find a proof in a book or online.

Summary

❏ Let $D \subseteq \mathbb{C}$ be a open set, and $f : D \to \mathbb{C}$ be a differentiable complex function. Let γ be a closed contour such that γ and its interior points lie in D.

Then, $\int_{\gamma} f = 0$.

CHAPTER 12

Cauchy's Integral Formula

An important consequence of Cauchy's Theorem is the surprising result that the value of a differentiable function at a point can be calculated by integrating over *any* contour around that point.

Theorem 12.1 (Cauchy's Integral Formula)
Suppose that $f : D \to \mathbb{C}$ is differentiable on the open set $D \subseteq \mathbb{C}$ and γ is a closed contour such that γ^* and its interior points are in D.

For all $w \in D \backslash \gamma^*$ we have

$$\int_\gamma \frac{f(z)}{z - w}\, dz = 2\pi i\, n(\gamma, w) f(w).$$

Proof. If $n(\gamma, w) = 0$, then w is not in the interior of γ and so $f(z)/(z - w)$ is differentiable on the interior of γ and the statement follows directly from Cauchy's Theorem. Without loss of generality we can assume that $n(\gamma, w) > 0$.

[Step 1] We shall first show that

$$\lim_{r \to 0} \int_{C_r} \frac{f(z) - f(w)}{z - w}\, dz = 0$$

where C_r is the circular contour $C_r(t) = w + re^{it}$, $(0 \le t \le 2\pi)$. For $r > 0$ sufficiently small C_r is contained in the interior of γ.

We have,

$$\left| \int_{C_r} \frac{f(z) - f(w)}{z - w} \, dz \right| \leq \sup_{z \in C_r^*} \left\{ \left| \frac{f(z) - f(w)}{z - w} \right| \right\} \times 2\pi r, \text{ by the Estimation Lemma,}$$

$$= \sup_{|z-w|=r} \left\{ \frac{|f(z) - f(w)|}{r} \right\} \times 2\pi r$$

$$= \sup_{|z-w|=r} \left\{ |f(z) - f(w)| \right\} \times 2\pi.$$

As $r \to 0$, we have $z \to w$ and as f is continuous, that $f(z) \to f(w)$, so

$$\sup_{|z-w|=r} \left\{ |f(z) - f(w)| \right\} \to 0 \text{ as } r \to 0.$$

Thus,

$$\lim_{r \to 0} \left| \int_{C_r} \frac{f(z) - f(w)}{z - w} \, dz \right| = 0.$$

As the limit is zero the same statement with the modulus signs removed is true.

[Step 2] Let β be a contour in $D \backslash \{w\}$ from the start point of γ to the start point of C_r.

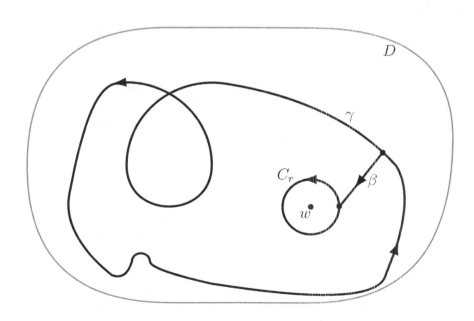

The contour

$$\widetilde{\gamma} = \gamma + \beta + \underbrace{(-C_r) + \cdots + (-C_r)}_{n(\gamma, w) \text{ times}} + (-\beta),$$

where there are $n(\gamma, w)$ copies of $-C_r$, has winding number zero about w, so by Cauchy's theorem applied to $\dfrac{f(z) - f(w)}{z - w}$ on $D\backslash\{w\}$,

$$\int_{\widetilde{\gamma}} \frac{f(z) - f(w)}{z - w} \, dz = 0$$

$$\left(\int_{\gamma} + \int_{\beta} - \int_{C_r} - \cdots - \int_{C_r} + \int_{-\beta}\right) \frac{f(z) - f(w)}{z - w} \, dz = 0$$

$$\int_{\gamma} \frac{f(z) - f(w)}{z - w} \, dz = n(\gamma, w) \int_{C_r} \frac{f(z) - f(w)}{z - w} \, dz.$$

Taking the limit as $r \to 0$ and using Step 1 shows that the left hand side is zero as is it does not depend on r. Therefore,

$$\int_{\gamma} \frac{f(z)}{z - w} \, dz = \int_{\gamma} \frac{f(w)}{z - w} \, dz = f(w) \int_{\gamma} \frac{1}{z - w} \, dz = 2\pi i n(\gamma, w) f(w).$$

$\square$

Remarks 12.2

(i) The theorem is remarkable and contrasts sharply with real analysis since if w is in the interior of γ, then

$$f(w) = \frac{1}{2\pi i \, n(\gamma, w)} \int_{\gamma} \frac{f(z)}{z - w} \, dz.$$

This says that *all* the values of f inside γ are completely determined by those on γ!

This behaviour is sometimes called 'action at a distance' because the values of f away from w determine the value $f(w)$.

(ii) We can recover the winding number if we use $f(z) = 1$ for all $z \in D$. (Although, to be honest, this is not surprising as we use this winding number result in the proof.)

(iii) The proof relies on Cauchy's Theorem so technically we have only proved the theorem for contours consisting of a finite number of lines and arcs. The full theorem follows from the fact that we prove Cauchy's theorem in more generality in Appendix A.

Examples 12.3

(i) Calculate $\int_\gamma \dfrac{\sin z}{4z - \pi} \, dz$, where $\gamma(t) = 3e^{-it}$, $0 \le t \le 4\pi$.

We can apply Cauchy's Integral Formula with

$$f(z) = \sin z \text{ and } w = \pi/4$$

(because we want $z - a$ for some a in the denominator, and if we take out the 4 we get $4(z - \pi/4)$).

The point $\pi/4$ lies within the circle formed by γ, and $n(\gamma, \pi/4) = -2$, because γ winds round $\pi/4$ clockwise twice. (Draw a picture!)

Hence,

$$
\begin{aligned}
\int_\gamma \frac{\sin z}{4z - \pi} \, dz &= \int_\gamma \frac{\sin z}{4(z - \pi/4)} \, dz \\
&= \frac{1}{4} \int_\gamma \frac{\sin z}{z - \pi/4} \, dz \\
&= \frac{1}{4} \times 2\pi i \times (-2) \times \sin(\pi/4) \\
&= -\pi i \times \frac{1}{\sqrt{2}} \\
&= -\frac{\pi i}{\sqrt{2}}.
\end{aligned}
$$

We could have applied Cauchy's Integral Formula with $f(z) = \dfrac{1}{4} \sin z$ and produced the same answer.

(ii) Let γ be a circle of radius 4 about 2, i.e., z such that $|z - 2| = 4$. Calculate

$$\int_\gamma \frac{e^z}{z^2 - 25} \, dz.$$

We have,

$$
\begin{aligned}
\int_\gamma \frac{e^z}{z^2 - 25} \, dz &= \int_\gamma \frac{1}{10} \frac{e^z}{z - 5} - \frac{1}{10} \frac{e^z}{z + 5} \, dz \\
&= \frac{1}{10} \int_\gamma \frac{e^z}{z - 5} \, dz - \frac{1}{10} \int_\gamma \frac{e^z}{z + 5} \, dz \\
&= \frac{1}{10} 2\pi i e^5 - 0, \text{ by Theorem 12.1,} \\
&= \frac{i \pi e^5}{10}.
\end{aligned}
$$

We shall use this consequence of Cauchy's Theorem to prove a number of surprising facts. The first is that if two differentiable functions are equal on a closed contour, then they are equal on its interior. This another example of 'action at a distance'.

Theorem 12.4

Suppose f and g are differentiable on an open set D and γ is a contour with its interior and γ^* in D. If $f(z) = g(z)$ for all $z \in \gamma^*$, then $f(z) = g(z)$ for all $z \in \mathrm{Int}(\gamma)$.

Proof. [We want to show that two objects are equal. A mathematician would think of doing this by showing their difference is 0.]

Let $w \in \mathrm{Int}(\gamma)$. Then by Cauchy's Integral Formula we have

$$
\begin{aligned}
f(w) - g(w) &= \frac{1}{2\pi i n(\gamma, w)} \int_\gamma \frac{f(z)}{z - w}\, dz - \frac{1}{2\pi i n(\gamma, w)} \int_\gamma \frac{g(z)}{z - w}\, dz \\
&= \frac{1}{2\pi i n(\gamma, w)} \int_\gamma \frac{f(z) - g(z)}{z - w}\, dz \\
&= \frac{1}{2\pi i n(\gamma, w)} \int_\gamma \frac{0}{z - w}\, dz, \quad \text{as } f(z) = g(z) \text{ for } z \in \gamma^*, \\
&= 0.
\end{aligned}
$$

$\square$

Exercises

Exercises 12.5

(i) Evaluate the following integral using Cauchy's Integral Formula:

(a) $\displaystyle \int_\gamma \frac{e^z + z}{z - 2}\, dz$ where γ is a circle of radius 1 centred at the origin,

(b) $\displaystyle \int_\gamma \frac{e^z + z}{z - 2}\, dz$ where γ is a circle of radius 3 centred at the origin,

(c) $\displaystyle \int_\gamma \frac{z^2}{z^2 + 1}\, dz$ where γ is the circle of radius 1 centred at i.

(d) $\displaystyle \int_\gamma |z + 2|^2\, dz$, where γ is a circle of radius 1 centred at the origin.

(e) $\displaystyle \int_{\gamma_1} \frac{z^2 + i}{z - 2}\, dz$, where $\gamma_1(t) = 3e^{2it}$, $(0 \le t \le 2\pi)$.

(f) $\int_{\gamma_2} \dfrac{z+2}{z(z+3i)}\, dz$, where $\gamma_2(t) = -3i + 2e^{-it}$, $(0 \le t \le 2\pi)$.

(g) $\int_{\gamma_3} \dfrac{z^4}{z-2}\, dz$, where $\gamma_3(t) = 3e^{-it}$, $(0 \le t \le 6\pi)$.

(h) $\int_{\gamma_4} \dfrac{z^3 - 18i}{z(z-6)}\, dz$, where $\gamma_4(t) = 3 + 4e^{it}$, $(0 \le t \le 4\pi)$.

(i) $\int_{\Gamma} \dfrac{\sinh z}{z - 3i}\, dz$ where $\Gamma = \alpha + \beta + \gamma - \beta$ and α, β, γ are the contours used in Exercises 5.18(iv).

(ii) In the following let γ be the contour given by $\gamma(t) = z_0 + 2e^{it}$ where $z_0 \in \mathbb{C}$ and $0 \le t \le 2\pi$.

    (a) Use Cauchy's Integral Formula to evaluate

$$\int_{\gamma} \frac{\cos(z)}{(z+1)^2(5z-4)}\, dz,$$

    where $z_0 = 3/2$.

    (b) Can we use Cauchy's Integral Formula to evaluate the above integral when $z_0 = i$? If so, then calculate the integral, if not, then give reasons.

    (c) Can we use Cauchy's Integral Formula to evaluate the above integral when $z_0 = 2i$? If so, then calculate the integral, if not, then give reasons.

(iii) Show that Cauchy's Theorem can be deduced from Cauchy's Integral Formula. (Hint: Consider the function $(z-w)f(z)$.)

(iv) Let D be an open set and f be differentiable on D. Show that for all points $z_0 \in D$ there exists $R > 0$ such that for all r with $0 < r < R$ we have

$$f(z_0) = \frac{1}{2\pi} \int_0^{2\pi} f(z_0 + re^{it})\, dt.$$

Note that this result says that the value of f at a point z_0 is the average of all values at points in a circle around z_0.

(v) Let $f : D \to \mathbb{C}$ be a complex differentiable function such that $\mathrm{Int}(\gamma_R) \subset D$ for some $R > 0$ where γ_R is the standard contour describing the circle of radius R based at the origin. Let $M = \sup_{z \in \gamma_R^*}\{|f(z)|\}$.

Show, for all w_1 and w_2 with $|w_1| < R$ and $|w_2| < R$, that

$$|f(w_1) - f(w_2)| \le \frac{|w_1 - w_2|RM}{(R - |w_1|)(R - |w_2|)}.$$

Summary

☐ Cauchy's Integral Formula: Let $D \subseteq \mathbb{C}$ be a open set and $f : D \to \mathbb{C}$ be differentiable. Let γ be a closed contour such that γ and its interior points are in D. If $w \in D \backslash \gamma^*$, then

$$\int_\gamma \frac{f(z)}{z - w} \, dz = 2\pi i \, n(\gamma, w) f(w).$$

☐ Cauchy's Integral Formula says that the value of a function at w is determined by the values on any contour round w.

CHAPTER 13

Taylor's Theorem

A differentiable real function need not be infinitely differentiable (see Example 13.8 below) and an infinitely differentiable real function need not be equal to its Taylor series (see Example 13.7). In this chapter we show, using Cauchy's Theorem, that the complex case is fundamentally different and significantly simpler. A differentiable complex function *is* infinitely differentiable and *is* equal to its Taylor series.

First, we need a definition so that we can determine the radius of convergence of a Taylor series.

Definition 13.1
Let S be a subset of $\mathbb{C}$ and w a point in $\mathbb{C}$. We define the **distance** from w to S, denoted $\mathrm{dist}(w, S)$, to be

$$\mathrm{dist}(w, S) = \inf\{|z - w| \,:\, z \in S\}.$$

We define $\mathrm{dist}(w, \emptyset) = \infty$.

In other words, dist is the shortest distance from w to a point of S.

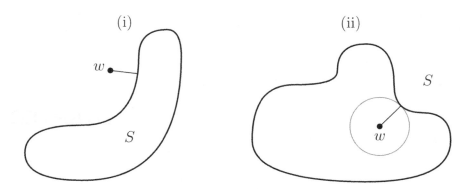

Note from diagram (ii) that the distance gives the radius for the largest circle centred at w whose interior lies in $\mathbb{C}\backslash S$.

Examples 13.2

  (i) Let $S = \{4 + i, 7 - 12i\}$. Then $\mathrm{dist}(3, S) = \sqrt{2}$.

  (ii) Let $S = \mathbb{C}\backslash\{z \,:\, |z| < 1\}$. Then $\mathrm{dist}(1/3, S) = 2/3$.

  (iii) Let $z_0 = 2i$. Then

      (a) $\mathrm{dist}(z_0, \mathbb{C}\backslash D) = 2$ where $D = \mathbb{C}\backslash\{0\}$,

      (b) $\mathrm{dist}(z_0, \mathbb{C}\backslash D) = 1$ where $D = \mathbb{C}\backslash\{1 + i\}$,

      (c) $\mathrm{dist}(z_0, \mathbb{C}\backslash D) = \infty$ where $D = \mathbb{C}$,

    The next theorem tells us that a complex differentiable function is equal to its Taylor series. This is a crucial difference between real and complex analysis and is the foundation for many applications of complex analysis.

Theorem 13.3 (Taylor's theorem for complex functions)
Suppose that $f : D \to \mathbb{C}$ is a differentiable function. Let $z_0 \in D$ and $R = \mathrm{dist}(z_0, \mathbb{C}\backslash D)$. Then, there exists a Taylor expansion of f about z_0, i.e., there exist $a_n \in \mathbb{C}$ such that

$$f(z) = \sum_{n=0}^{\infty} a_n(z - z_0)^n, \text{ for all } |z - z_0| < R.$$

Furthermore,

$$a_n = \frac{f^{(n)}(z_0)}{n!} = \frac{1}{2\pi i} \int_{C_r} \frac{f(z)}{(z - z_0)^{n+1}} \, dz,$$

where C_r is a circle of radius r about z_0 and r is any number with $0 < r < R$.

The expression for a_n is sometimes known as **Cauchy's formula for derivatives**.

Proof. Without loss of generality we can assume that $z_0 = 0$.

Let $z \in \text{Int}(C_r)$ where C_r is centred at z_0 with radius r. Then $|z| < r$, i.e., $|z| < |\zeta| = r$ for $\zeta \in C_r^*$.

By Cauchy's Integral Formula we have

$$f(z) = \frac{1}{2\pi i} \int_{C_r} \frac{f(\zeta)}{\zeta - z} \, d\zeta.$$

Consider the integrand, we expand it as a series:

$$\frac{f(\zeta)}{\zeta - z} = \frac{f(\zeta)}{\zeta \left(1 - \frac{z}{\zeta}\right)}$$

$$= \frac{f(\zeta)}{\zeta} \sum_{n=0}^{\infty} \left(\frac{z}{\zeta}\right)^n, \text{ as } \left|\frac{z}{\zeta}\right| < 1 \text{ and } \sum x^n = \frac{1}{1-x},$$

$$= \sum_{n=0}^{\infty} \frac{f(\zeta)z^n}{\zeta^{n+1}}.$$

[We want to integrate the left hand side of the equation. We can do this by term-by-term integration of the right hand side. To ensure we are allowed to do this we shall use Theorem 9.10.]

We now show that the series satisfies conditions of Theorem 9.10. Let

$$g(\zeta) = \frac{f(\zeta)}{\zeta - z} \text{ and } g_n(\zeta) = \frac{f(\zeta)z^n}{\zeta^{n+1}}$$

Then define $M_n = \sup_{\zeta \in C_r^*} \{|g_n(\zeta)|\}$. We see that

$$M_n = \sup_{|\zeta|=r} \left\{ \left|\frac{f(\zeta)z^n}{\zeta^{n+1}}\right| \right\} = \sup_{|\zeta|=r} \left\{ \frac{|f(\zeta)||z|^n}{r^{n+1}} \right\} = \left(\sup_{|\zeta|=r} \{|f(\zeta)|\}\right) \frac{|z|^n}{r^{n+1}}.$$

The supremum exists. (This is because $|f(\zeta)|$ is a continuous real function on $[0, 2\pi]$ as $\zeta = re^{it}$ for $t \in [0, 2\pi]$ and a continuous function on an interval attains its bounds.)

By the ratio test and $|z| < r$ we have that $\sum M_n$ converges. (Exercise!) Obviously, by definition $|g_n(\zeta)| \leq M_n$ for all $\zeta \in C_r^*$ so by Corollary 9.10 we can exchange $\sum$ and $\int$ between lines 2 and 3 in the following argument:

$$
\begin{aligned}
f(z) &= \frac{1}{2\pi i} \int_{C_r} \frac{f(\zeta)}{\zeta - z} \, d\zeta \\
&= \frac{1}{2\pi i} \int_{C_r} \sum_{n=0}^{\infty} \frac{f(\zeta) z^n}{\zeta^{n+1}} \, d\zeta \\
&= \frac{1}{2\pi i} \sum_{n=0}^{\infty} \int_{C_r} \frac{f(\zeta) z^n}{\zeta^{n+1}} \, d\zeta \\
&= \sum_{n=0}^{\infty} \left(\frac{1}{2\pi i} \int_{C_r} \frac{f(\zeta)}{\zeta^{n+1}} \, d\zeta \right) z^n \\
&= \sum_{n=0}^{\infty} a_n z^n.
\end{aligned}
$$

Note that $a_n = \dfrac{1}{2\pi i} \displaystyle\int_{C_r} \frac{f(\zeta)}{\zeta^{n+1}} \, d\zeta$ does not depend on z and because of this the result holds for $z \in \text{Int}(C_R)$.

That $a_n = \dfrac{f^{(n)}(0)}{n!}$ follows from Corollary 9.6. $\qquad\qquad\square$

Let's see the theorem in action.

Example 13.4

(i) Find the Taylor series of $f(z) = (2z - 5i)^3$ at $z = -2i$. We have,

$$
\begin{aligned}
f(z) &= (2z - 5i)^3 &\implies& \quad f(4i) &= (2z - 5i)^3 = (3i)^3 = -27i, \\
f'(z) &= 12(2z - 5i)^2 &\implies& \quad f'(4i) &= 12(3i)^2 = -54, \\
f''(z) &= 24(2z - 5i) &\implies& \quad f''(4i) &= 24(3i) = 72i, \\
f^{(3)}(z) &= 48 &\implies& \quad f^{(3)}(4i) &= 48, \\
f^{(4)}(z) &= 0 &\implies& \quad f^{(4)}(4i) &= 0.
\end{aligned}
$$

Obviously, $f^{(4)}(z) = 0$ implies that $f^{(n)})(z) = 0$ for all $n > 4$. So,

$$
\begin{aligned}
f(z) &= f(4i) + f'(4i)(z - 4i) + \frac{f''(4i)}{2!}(z - 4i)^2 + \frac{f^{(3)}(4i)}{3!}(z - 4i)^3 \\
&= -27i - 54(z - 4i) + \frac{72i}{2}(z - 4i)^2 + \frac{48}{6}(z - 4i)^3 \\
&= -27i - 54(z - 4i) + 36i(z - 4i)^2 + 8(z - 4i)^3.
\end{aligned}
$$

Note that this expansion is valid for all $z \in \mathbb{C}$ as f is differentiable on all of $\mathbb{C}$, i.e.,

$$
R = \text{dist}(4i, \mathbb{C}\backslash\mathbb{C}) = \text{dist}(4i, \emptyset) = \infty.
$$

(Check: The final function is a polynomial so is defined on all of $\mathbb{C}$.)

(ii) Expand $f(z) = \dfrac{1}{1-z}$ about 0. It is easy to show that $f^{(n)}(z) = \dfrac{n!}{(1-z)^{n+1}}$, so $f^{(n)}(0) = n!$.

Thus,

$$\frac{1}{1-z} = f(z) = \sum_{n=0}^{\infty} \frac{f^{(n)}(0)}{n!}(z-0)^n = \sum_{n=0}^{\infty} \frac{n!}{n!}z^n = \sum_{n=0}^{\infty} z^n.$$

This is valid for all $|z| < 1$ because f is differentiable on $\mathbb{C}\backslash\{1\}$ and so

$$R = \text{dist}(0, \mathbb{C}\backslash(\mathbb{C}\backslash\{1\})) = \text{dist}(0, \{1\}) = 1.$$

Note that this just confirms a result we already know (which is in fact used in the proof of Taylor's Theorem anyway so is not surprising).

The 'Furthermore' part of Taylor's Theorem is often overlooked. In one sense it is saying that we can calculate the coefficients of the Taylor series using integrals. On the other hand, we can also see it as allowing us to calculate integrals using the coefficients. The key part of this is that Taylor series coefficients are calculated using derivatives. In other words, certain integrals (hard to calculate) can be calculated using derivatives (easy to calculate). This is surely a great insight.

Example 13.5

Evaluate the integral

$$\int_{C_r} \frac{ze^{2z}}{z(z-6i)-9}\, dz \text{ where } r > 0.$$

Solution: [The key is to note that $z(z-6i) - 9 = z^2 - 6i - 9 = (z-3i)^2$.] If we let $f(z) = ze^{2z}$, then by Taylor's Theorem

$$f'(3i) = \frac{1}{2\pi i}\int_{C_r} \frac{f(z)}{(z-3i)^2}\, dz.$$

Then, $f'(z) = e^{2z}(2z+1)$ so $f'(3i) = (1+6i)e^{6i}$ and

$$\int_{C_r} \frac{ze^{2z}}{z(z-6i)-9}\, dz = 2\pi i\left((1+6i)e^{6i}\right) = 2\pi(i-6)e^{6i}.$$

Complex differentiable and analytic are the same

Definition 13.6
A real or complex function equal to its Taylor series is called **analytic**.

So for example, sine, cosine and the exponential function are all analytic. (In the case of the complex versions that is in fact how we defined them!)

In *real analysis* we have the following implications about functions:

$$\text{analytic} \implies \text{infinitely differentiable} \implies \text{differentiable.} \tag{13.1}$$

The converse of each implication is not true. First, infinitely differentiable does not imply analytic as the next example shows.

Example 13.7
Let $f : \mathbb{R} \to \mathbb{R}$ be defined by

$$f(x) = \begin{cases} e^{-1/x} & x > 0 \\ 0 & x \leq 0. \end{cases}$$

It can be shown (see exercises) that $f^{(n)}(0) = 0$ for all $n > 0$, i.e., all the higher derivatives of f at 0 are 0. Thus the Taylor series at 0 for f is the zero function. Clearly, f is not zero on the positive axis near 0 and hence f, despite being infinitely differentiable, is not equal to its Taylor series.

The converse of the second implication in (13.1), i.e., that infinitely differentiable implies differentiable, is false for real functions.

Example 13.8
Suppose that $f : \mathbb{R} \to \mathbb{R}$ is defined by

$$f(x) = \begin{cases} x^2 & x \geq 0 \\ 0 & x < 0. \end{cases}$$

Then, f is differentiable with derivative

$$f'(x) = \begin{cases} 2x & x \geq 0 \\ 0 & x < 0. \end{cases}$$

This derivative is not differentiable at 0, so f is not infinitely differentiable.

A major surprise in complex analysis is that both the implications above can be reversed. Their converses *are* true for complex functions. That is, in complex analysis we have the following implications, indeed equivalences, for functions:

$$\text{analytic} \iff \text{infinitely differentiable} \iff \text{differentiable}. \qquad (13.2)$$

First, if the function is analytic, then by definition it can be be given as a power series, and so is infinitely differentiable by Corollary 9.5. If a function is infinitely differentiable then, trivially, it is differentiable. These two facts show that (13.1) is true for complex functions. Next, Taylor's Theorem says that for complex differentiable function f there exists an open disc upon which f is equal to its power series expansion. That is, differentiable implies analytic. Hence, (13.2) is true.

The important conclusion of this is that

differentiable and analytic are equivalent in complex analysis.

This explains why some authors use the word analytic to mean complex differentiable.

How to think like a mathematician 13.9
When solving problems concerning differentiable complex functions, whether in calculations or in theoretical ones, use the fact that the function can be written locally as a power series.

Taylor series of a product

Theorem 13.10
Suppose that $f(z) = \sum_{n=0}^{\infty} a_n z^n$ for all $|z| < R_1$ and $g(z) = \sum_{n=0}^{\infty} b_n z^n$ for all $|z| < R_2$. Then, $f(z)g(z) = (fg)(z) = \sum_{n=0}^{\infty} c_n z^n$ for all $|z| < \min\{R_1, R_2\}$, where $c_n = \sum_{k=0}^{n} a_k b_{n-k}$.

Proof. Let $R = \min\{R_1, R_2\}$. Then, f and g are complex differentiable for all $|z| < R$. The product of two differentiable functions is differentiable by the product rule. So, fg is differentiable. By Theorem 13.3 it has a series expansion for $|z| < R$, and by Corollary 9.6, we get

$$(fg)(z) = \sum_{n=0}^{\infty} \frac{(fg)^{(n)}(0)}{n!} z^n.$$

Now we can apply Leibniz's rule

$$\begin{aligned}
(fg)^{(n)}(0) &= \sum_{k=0}^{n} \frac{n!}{k!(n-k)!} f^{(k)}(0) g^{(n-k)}(0) \\
&= \sum_{k=0}^{n} n! \frac{f^{(k)}(0)}{k!} \frac{g^{(n-k)}(0)}{(n-k)!} \\
&= n! \sum_{k=0}^{n} a_k b_{n-k}.
\end{aligned}$$

Therefore,

$$(fg)(z) = \sum_{n=0}^{\infty} c_n z^n, \quad (|z| < R), \quad \text{where } c_n = \sum_{k=0}^{n} a_k b_{n-k}.$$

$\square$

Corollary 13.11
Suppose f and g are power series with positive radii of convergence. If $g(0) \neq 0$, then f/g has a power series expansion at 0 with positive radius of convergence.

Proof. Define $h(z) = 1/g(z)$. By the differentiability of g and the fact that $g(0) \neq 0$, there is an open disc centred at 0 upon which h is differentiable. (Exercise.) Thus, h has a power series expansion, and the result follows from the previous theorem applied to fh. $\square$

Remark 13.12
The precise value of the radius of convergence will depend on where g is zero.

Example 13.13
Express

$$h(z) = \frac{e^{3z}}{1 - 2z}$$

as a power series about $z = 1$.

Solution: [Our initial thought may have been to find the derivatives of h and evaluate at 1. This is in fact feasible but the repeated use of the quotient rule makes the calculations messy. Instead we can view the function as a product.]

Let's find power series expansions for e^{3z} and $\dfrac{1}{1-2z}$ about $z = 1$ and multiply them together. We must find a_n and b_n such that

$$e^{3z} = f(z) = \sum_{n=0}^{\infty} a_n (z-1)^n \qquad \text{and} \qquad \frac{1}{1-2z} = g(z) = \sum_{n=0}^{\infty} b_n (z-1)^n.$$

Now,

$$
\begin{aligned}
f(z) &= e^{3z} \\
\implies f^{(n)}(z) &= 3^n e^{3z} \\
\implies f^{(n)}(1) &= 3^n e^3 \\
\implies a_n &= \frac{f^{(n)}(1)}{n!} = \frac{3^n e^3}{n!}.
\end{aligned}
$$

And,

$$
\begin{aligned}
g(z) &= \frac{1}{1 - 2z} \\
\implies g^{(n)}(z) &= \frac{2^n n!}{(1 - 2z)^{n+1}} \\
\implies g^{(n)}(1) &= \frac{2^n n!}{(-1)^{n+1}} \\
\implies b_n &= \frac{g^{(n)}(1)}{n!} = \frac{2^n}{(-1)^{n+1}} = (-1)^{n+1} 2^n.
\end{aligned}
$$

Hence, from Theorem 13.10 we have

$$
\frac{e^{3z}}{1 - 2z} = f(z)g(z) = \sum_{n=0}^{\infty} c_n (z - 1)^n
$$

where

$$
\begin{aligned}
c_n &= \sum_{k=0}^{n} a_k b_{n-k} = \sum_{k=0}^{n} \frac{3^k e^3}{k!} (-1)^{n-k+1} 2^{n-k} \\
&= -e^3 \sum_{k=0}^{n} (-1)^{n-k} \left(\frac{3}{2}\right)^k \frac{2^n}{k!} = -e^3 2^n \sum_{k=0}^{n} \frac{(-1)^{n-k}}{k!} \left(\frac{3}{2}\right)^k.
\end{aligned}
$$

Thus

$$
\frac{e^{3z}}{1 - 2z} = -e^3 \sum_{n=0}^{\infty} \left(\sum_{k=0}^{n} \frac{(-1)^{n-k}}{k!} \left(\frac{3}{2}\right)^k \right) 2^n (z - 1)^n.
$$

The radius of convergence of f is obviously infinite. The radius of convergence of g is $1/2$ since $1/(1 - 2z)$ is not defined at $z = 1/2$ and $1 - 1/2 = 1/2$. Thus the radius of convergence of the above series is $\min\{1/2, \infty\} = 1/2$.

The Identity Theorem

We now state a versatile result that has very weak assumptions but a very strong conclusion.

Theorem 13.14 (Identity Theorem)
Let D be a path-connected open set and f be a complex differentiable function on D. If there exists a sequence of distinct point $z_n \in D$ such that $z_n \to z_0 \in D$ and $f(z_n) = 0$ for all n, then $f(z) = 0$ for all $z \in D$.

Proof. [Step 1] First we prove for D an open disc centred at z_0. Assume that f is non-zero. Let a_m be the first non-zero coefficient in the Taylor's series of f, i.e.,

$$f(z) = a_m(z - z_0)^m + a_{m+1}(z - z_0)^{m+1} + a_{m+2}(z - z_0)^{m+2} + \dots$$

Let

$$g(z) = a_m + a_{m+1}(z - z_0) + a_{m+2}(z - z_0)^2 + \dots$$

Then, g converges on D (exercise) and hence is differentiable on D. Also, $g(z_n) = 0$ for all n. Hence,

$$0 = a_m + a_{m+1}(z_n - z_0) + a_{m+2}(z_n - z_0)^2 + \dots$$

for all n. As $n \to \infty$ we have $z_n \to z_0$ and so $a_m = 0$. This is a contradiction so $f(z) = 0$ for all z in the disc.

[Step 2] Let w be any point of D and γ be a non-intersecting contour in D starting at z_0 and ending at w. As D is open there exists an open disc around z_0 upon which f is identically zero. In particular one can find a closed interval that maps via γ such that f is zero on the image of γ. Now let $\widetilde{w} = \gamma(\widetilde{t})$ be the last point on γ^* such that $f(\gamma(t)) = 0$.

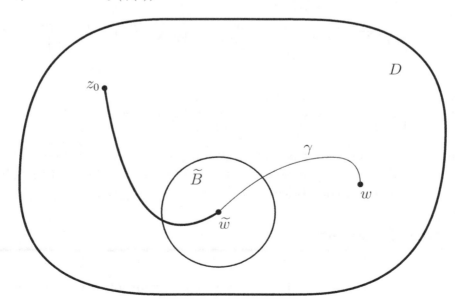

If $\widetilde{w} = w$, then we are done since $f(w) = 0$ and w was arbitrary. Assume that $w \neq \widetilde{w}$. There exists an open disc $\widetilde{B}$ around $\widetilde{w}$. Choosing points on $\gamma^* \cap \widetilde{B}$ with $t < \widetilde{t}$ we can produce a sequence z_n such that $z_n \to \widetilde{w}$ and $f(z_n) = 0$ for all n. By Step 1, $f(z) = 0$ for all $z \in \widetilde{B}$ and in particular $\widetilde{w}$ is not the last point in γ^* with $f(\gamma(t)) = 0$. This is a contradiction and finishes the proof. $\qquad\square$

Corollary 13.15

Let D be a path-connected open set and f and g be complex differentiable functions on D. If there exists a sequence of distinct points $z_n \in D$ such that $z_n \to z_0 \in D$ and $f(z_n) = g(z_n)$ for all n, then $f(z) = g(z)$ for all $z \in D$.

Proof. Apply the theorem to the function $f - g$. $\qquad\square$

We can use this theorem and its corollary to prove trigonometric identities. In fact, we can extend *all* real trigonometric identities to complex ones.

Example 13.16

We can show that $\cos(2z) = 1 - \sin^2(z)$:

Let $f(z) = \cos(2z)$ and $g(z) = 1 - \sin^2(z)$ for all $z \in \mathbb{C}$. Then $f(z) = g(z)$ for all $z \in \mathbb{R}$. In particular, there exists a sequence $z_n \in \mathbb{R}$ with limit in $\mathbb{R}$ giving $f(z_n) = g(z_n)$. (We can take $z_n = 1/n$.) By the corollary, $f(z) = g(z)$ for all $z \in \mathbb{C}$, i.e., $\cos(2z) = 1 - \sin^2(z)$.

Exercise 13.17

Prove that $\sin(2z) = 2\cos(z)\sin(z)$ for all $z \in \mathbb{C}$.

Remark 13.18

The corollary shows that there is really only one way to extend each of exp, sin and cos as functions from the real to complex plane so that the resulting complex functions are differentiable. The details are left as an exercise.

Morera's Theorem

There is a partial converse to Cauchy's theorem.

Theorem 13.19

Let $f : D \to \mathbb{C}$ be a continuous map on a path-connected open set D. If $\int_\gamma f = 0$ for all closed contours γ in D, then f is analytic.

Proof. By Exercise 8.29(xxiii) there exists $F : D \to \mathbb{C}$ such that $F' = f$. But the derivative of a differentiable function is itself differentiable since differentiable complex functions are analytic. $\qquad\square$

Exercises

Exercises 13.20

(i) Find the Taylor series at the indicated points and the domain upon which it is equal to the function:

   (a) $(z + i)^4$ at $z = -2$,

   (b) $\dfrac{z^2}{(1 - z)^2}$ at $z = 0$,

   (c) e^{z^2} at $z = 0$,

   (d) $\cos z$ at $z = 3\pi/4$.

(ii) Prove that $\sum_{n=0}^{\infty} M_n$ in the proof of Theorem 13.3 converges.

(iii) Prove that if $f(z) = f'(z)$ for all $z \in \mathbb{C}$, then $f(z) = Ce^z$ for some $C \in \mathbb{C}$.

(iv) Show that

$$\text{Log}(z) = \left(\sum_{n=1}^{\infty} (-1)^{n-1} \frac{(z-1)^n}{n} \right) \quad \text{for } |z - 1| < 1.$$

(v) Let g be analytic on the open set D with $g(z_0) \neq 0$ and let f be defined by $f(z) = (z - z_0)^m g(z)$. Show that if γ is a contour with $z_0 \notin \gamma^*$, then

$$\frac{1}{2\pi i} \int_{\gamma} \frac{f'(z)}{f(z)} \, dz = n(\gamma, z_0)m$$

(Hint: You will need an exercise from Chapter 10.)

(vi) Suppose that f is analytic on $\mathbb{C}$ and that there exists $M \in \mathbb{R}$ and $n \in \mathbb{N}$ such that $|f(z)| \leq M|z|^n$ for all $z \in \mathbb{C}$. Show that f is a polynomial. What is its maximum degree?

(vii) Evaluate the following integrals.

   (a) $\displaystyle \int_{\gamma} \frac{z^4 + 2z - 9}{z^3 - 6z^2 + 12z - 8} \, dz$ where γ is the unit circle about 2,

   (b) $\displaystyle \int_{\gamma} \frac{(z - i)^2 + (z - 1)^3}{(z - 1)(z - i)^2} \, dz$ where γ is the circle of radius 4 about i.

(viii) Consider Corollary 13.11.

(a) Prove the fact needed in the proof: For $h(z) = 1/g(z)$ with g differentiable and non-zero at 0 there is an open disc centred at 0 upon which h is differentiable.

(b) Make the statement of the corollary more precise. That is, what is the radius of convergence?

(ix) Suppose that f is defined by $f(z) = \dfrac{1}{1 - 3z}$.

(a) Show that $f^{(k)}(z) = \dfrac{3^k k!}{(1 - 3z)^{k+1}}$.

(b) Write down the Taylor series for f at 0.

(c) Find the radius of convergence for this series.

(x) Show that the Identity Theorem does not hold for real functions.

(xi) Prove that all real trigonometric identities extend to complex trigonometric identities.

(xii) Prove that the **Binomial Theorem** holds for complex numbers. That is

$$(w + z)^n = \sum_{k=0}^{n} \frac{n!}{(n - k)!k!} w^{n-k} z^k$$

for all $z, w \in \mathbb{C}$. (Hint: Prove first for $w = 1$.)

(xiii) Let D be a path-connected open set and γ a contour such that $\gamma^* \subset D$. Let $f : D \to \mathbb{C}$ be a continuous complex function. Show that if f is complex differentiable on $D \backslash \gamma^*$, then f is differentiable on D.

(xiv) Show that the function g in the proof of Theorem 13.14 is convergent.

(xv) Show that the extensions to the complex plane of sin, cos and exp are unique if we require the extensions to be differentiable.

(xvi) The nth Legendre polynomial, p_n, is defined to be

$$p_n(z) = \frac{1}{n!2^n} \frac{d^n}{dz^n} \left((z^2 - 1)^n \right).$$

(These polynomials are useful in solving Laplace's equation in spherical coordinates.) Show that

$$p_n(z) = \frac{1}{2^{n+1}\pi i} \int_\gamma \frac{(\zeta^2 - 1)^n}{(\zeta - z)^{n+1}} \, d\zeta$$

where γ is a circle around z.

(xvii) We need to show that the function f in Example 13.7 has $f^{(n)}(0) = 0$ for all n.

   (a) Inductively define the polynomials p_n by $p_{n+1}(t) = t^2(p_n(t) - p'_n(t))$ with $p_0(t) = 1$. Prove that for all $x > 0$

   $$f^{(n)}(x) = e^{-1/x} p_n \left(\frac{1}{x} \right).$$

   (b) Show that $f^{(n)}(0) = \lim_{x \to 0^+} e^{-1/x} p_n \left(\frac{1}{x} \right)$.

   (c) Use the substitution $t = 1/x$ to show that $f^{(n)}(0) = 0$ for all n.

Summary

❏ If a complex function is differentiable on a domain, then at every point there is a power series expansion valid on some neighbourhood.

❏ Differentiable complex functions are infinitely differentiable.

❏ If S_1 and S_2 are power series with radii of convergence R_1 and R_2, then $S_1 S_2$ is a power series with radius of convergence $\min\{S_1, S_2\}$.

❏ If S is a power series at z_0 with positive radius of convergence, then $1/S$ has a power series expansion at z_0 provided $S(z_0) \neq 0$.

❏ There is a partial converse to Cauchy's Theorem.

CHAPTER 14

Surprising Consequences of Cauchy's Theorem

From Cauchy's Theorem (via Cauchy's Integral Formula) we will deduce some surprising consequences. These are surprising in the sense that there are no analogous results in real analysis.

Fundamental Theorem of Algebra

The remarkable part of the next theorem is not the statement, you probably already know it. What is noteworthy is that despite being called the **Fundamental Theorem of Algebra** it is really a theorem of analysis not algebra.

Theorem 14.1 (Fundamental Theorem of Algebra)
Every non-constant polynomial has a root in $\mathbb{C}$.

Proof. Suppose not and derive a contradiction. Let $p(z) = a_n z^n + a_{n-1} z^{n-1} + \ldots a_1 z + a_0$, $n \geq 1$, $a_n \neq 0$. Since $p(z) \neq 0$ for all $z \in \mathbb{C}$ the function defined by $1/p(z)$ is differentiable on all of $\mathbb{C}$. Now, for $z \neq 0$,

$$\left| \frac{p(z)}{z^n} \right| = \left| a_n + \frac{a_{n-1}}{z} + \cdots + \frac{a_0}{z^n} \right| \to |a_n| \text{ as } |z| \to \infty.$$

So there exists an R such that $|z| \geq R$ implies that $\left| \dfrac{p(z)}{z^n} \right| \geq \dfrac{|a_n|}{2}$. Thus $\left| \dfrac{1}{p(z)} \right| \leq \dfrac{2}{|a_n||z|^n}$ for such z.

As $1/p(z)$ is differentiable on all of $\mathbb{C}$ we have for all r, by Cauchy's Integral Formula,

$$\left| \frac{1}{p(0)} \right| = \left| \frac{1}{2\pi i} \int_{C_r} \frac{1/p(z)}{z - 0} \, dz \right|$$
$$\leq \frac{1}{2\pi} \frac{2 \times 2\pi r}{|a_n| r^n \times r}, \text{ by the Estimation Lemma and for } r > R.$$

The right hand side can be made as small as we like by taking r large enough. Thus $1/p(0) = 0$. This is impossible so we have a contradiction and therefore p has a root. $\qquad\square$

Liouville's Theorem

Definition 14.2
A complex function $f : D \to \mathbb{C}$ is called **bounded** if there exists a real number M such that $|f(x)| \leq M$ for all $z \in D$.

The next theorem is rather unexpected.

Theorem 14.3
Suppose that f is differentiable and bounded on all of $\mathbb{C}$. Then, f is constant.

Proof. Let z_0 be an arbitrary point in $\mathbb{C}$ and let C_r be the contour giving a circle of radius r around it. Then, by the description of derivatives in Theorem 13.3, we have

$$|f'(z_0)| = \frac{1}{2\pi} \left| \int_{C_r} \frac{f(z)}{(z - z_0)^2} \, dz \right|$$
$$\leq \frac{1}{2\pi} \sup_{|z - z_0| = r} \left| \frac{f(z)}{(z - z_0)^2} \right| \times 2\pi r, \text{ by the Estimation Lemma,}$$
$$= \sup_{|z - z_0| = r} \frac{|f(z)|}{r^2} \times r$$
$$\leq \frac{M}{r}.$$

But r was arbitrary and as r goes to infinity we see that $f'(z_0) = 0$ (since it does not depend on r). Since z_0 was also arbitrary, we deduce from Theorem 8.15 that f is constant. □

Remark 14.4

Contrast this with real analysis: Bounded does not imply constant. For example, consider the sine function. This is differentiable on all of $\mathbb{R}$ and $|\sin x| \leq 1$ for all $x \in \mathbb{R}$ but sin is not constant.

Exercise 14.5

Show, using Liouville's Theorem, that cos and sin are not bounded on $\mathbb{C}$.

Exercise 14.6

In the proof of Theorem 14.3 we showed that the modulus of the derivative of a function was bounded by an expression involving r. Generalize this to higher derivatives. (These are called Cauchy's Inequalities.)

Local Maximum Modulus Theorem

We cannot order complex numbers and hence to define local maxima of functions we should look at their modulus.

Definition 14.7

Let $f : D \to \mathbb{C}$ be a complex function on an open set D. We say f has a **local maximum at** z_0 if there exists an open disc $B \subseteq D$ such that $|f(z)| \leq |f(z_0)|$ for all $z \in B$.

In real analysis an important skill is the ability to find local maxima and minima. For complex analysis the next theorem tells us that searching for local maxima is pointless.

Theorem 14.8 (Local Maximum Modulus Theorem)

Suppose that $f : D \to \mathbb{C}$ is a complex differentiable function and D is path-connected. If f has a local maximum in D, then f is constant on D.

Proof. Suppose the local maximum is at z_0. By Exercise 12.5(iv) we have

$$|f(z_0)| \leq \left| \frac{1}{2\pi} \int_0^{2\pi} f(z_0 + re^{it})\, dt \right| \leq \frac{1}{2\pi} \int_0^{2\pi} \left| f(z_0 + re^{it}) \right| dt$$

for all $0 < r < R$, for some R. From this and $|f(z_0)| \geq |f(z_0 + re^{it})|$ we have

$$0 \leq \frac{1}{2\pi} \int_0^{2\pi} |f(z_0)| - \left| f(z_0 + re^{it}) \right| dt \leq 0.$$

The integrand is a non-negative continuous function and the integral is zero so the integrand must be zero too. Thus

$$|f(z_0)| = |f(z_0 + re^{it})|$$

for all $0 \le r \le R$ and $0 \le t \le 2\pi$. So $|f|$ is constant on an open disc, B say. By Exercise 8.29(xxi) we see that f is constant on B.

Now define the constant function $g(z) = f(z_0)$ for all $z \in D$. As f is constant on B we can construct a sequence z_n with $z_n \to z_0$ such that $f(z_n) = f(z_n)$ for all n. Hence, by Corollary 13.15 we have $g = f$ on D and hence f is constant on D. □

Exercise 14.9
Define local minima. Conjecture and prove for local minima a theorem analogous to the Local Maximum Modulus Theorem.

Maximum Modulus Principle

Theorem 14.10
Let $f : D \to \mathbb{C}$ be a differentiable complex function and γ be a closed contour such that $\gamma^* \cup Int(\gamma) \subset D$.

If $|f(z)| \le M$ for all $z \in \gamma^*$, then $|f(w)| \le M$ for all $w \in Int(\gamma)$.

Remark 14.11
The theorem says that the modulus of a function within the interior of a contour is never bigger than the modulus of the function on the contour. In other words the maximum modulus of a function always occurs on the boundary of a region.

Proof. This proof contains a fancy trick. We shall apply Cauchy's Integral Formula to $w \in Int(\gamma)$ and $f(z)^k$, where k is a natural number:

$$f(w)^k = \frac{1}{2\pi i\, n(\gamma, w)} \int_\gamma \frac{f(z)^k}{z - w}\, dz.$$

Obviously $|z - w| \ge dist(w, \gamma^*)$ for all $z \in \gamma^*$.

So,

$$|f(w)|^k \le \frac{1}{2\pi\, |n(\gamma, w)|} \frac{M^k}{dist(w, \gamma^*)} L(\gamma), \text{ by the Estimation Lemma,}$$

since

$$\frac{|f(w)|^k}{|z - w|} \le \frac{M^k}{dist(w, \gamma^*)} \text{ for } z \in \gamma^*.$$

Therefore,

$$|f(w)| \leq \left(\frac{L(\gamma)}{2\pi \, |n(\gamma, w)| \, \text{dist}(w, \gamma^*)} \right)^{1/k} M.$$

Now, $\lim_{k \to \infty} x^{1/k} = 1$ for all $x > 0$ as follows: For $x \in \mathbb{R}$ with $x > 0$,

$$
\begin{aligned}
x^{1/k} &= e^{(1/k) \ln x} \\
&\to e^0 \text{ as } k \to \infty \\
&= 1.
\end{aligned}
$$

Thus, letting $k \to \infty$ gives $|f(w)| \leq M$. $\qquad \square$

Exercise 14.12
Use the maximum modulus principle to prove Theorem 12.4.

The 'furthermore' in the following corollary shows how rigid complex differentiable function are. That is weak assumptions can lead to strong conclusions. Here's another example of this.

Corollary 14.13 (Schwarz Lemma)
Let $D = \{z \in \mathbb{C} : |z| < R\}$ and $f : D \to D$ be a differentiable function with $f(0) = 0$. Then, $|f'(0)| \leq 1$ and $|f(z)| \leq |z|$ for all $z \in D$.

Furthermore, if $|f(z)| = |z|$ for any point $z \in D \backslash \{0\}$, then $f(z) = cz$ for some $c \in \mathbb{C}$ with $|c| = 1$.

Proof. Define $g : D \to D$ by

$$g(z) = \begin{cases} \dfrac{f(z)}{z}, & z \neq 0 \\ f'(0), & z = 0. \end{cases}$$

This is differentiable on D. (Exercise.) Let C_r be the contour giving the circle of radius r centred at the origin, with $r < R$. Then, by the Maximum Modulus Principle there exists $z^* \in (C_r)^*$ such that for all $z \in D$,

$$|g(z)| \leq |g(z^*)| = \left| \frac{f(z^*)}{z^*} \right| \leq \frac{R}{r}.$$

Taking the limit as $r \to R$ we see that $|g(z)| \leq 1$ which proves the two facts in the first part of the statement.

For the second part, notice that if $|g(z)| = 1$ for some z in $D \backslash \{0\}$, then this g has a local maximum at z. By the Local Maximum Modulus Theorem this implies that g is constant, so $g(z) = c$ say for all z. Hence, $f(z) = cz$. $\qquad \square$

Exercises

Exercises 14.14

(i) Let p be a non-constant polynomial. Show that

$$p(z) = (z - \alpha_1)(z - \alpha_2) \ldots (z - \alpha_n)$$

for some n and $\alpha_1, \alpha_2, \ldots, \alpha_n \in \mathbb{C}$.

(ii) Generalize Liouville's Theorem by replacing the bounded condition with the condition that $f(z)/z \to 0$ as $|z| \to \infty$.

(iii) Suppose that f is complex differentiable on all of $\mathbb{C}$ and $\text{Re}(f(z)) \leq \text{Im}(f(z))$ for all $z \in \mathbb{C}$. Use the function $g(z) = \exp\left((1 + i)f(z)\right)$ to show that f is constant.

(iv) Prove that the function g in the proof of Corollary 14.13 is differentiable on all of D.

(v) Let f be an analytic function on the open disk D centered at the origin and of radius r. Suppose that $f(0) = 0$ and $|f(z)| < M$ for all $z \in D$.

   (a) Prove that $|f(z)| \leq \dfrac{r}{M}|z|$ for all $z \in D$.

   (b) Hence, deduce Liouville's Theorem.

   (c) Show that if $r = M$, then $|f'(0)| \leq 1$.

(vi) Use Exercise 12.5(v) to give an alternative proof of Liouville's Theorem.

(vii) Suppose that f is analytic on $\mathbb{C}$ and $|f(z)| \leq M|z|$ for all z and some $M \in \mathbb{R}$. Prove that f is a linear function. [Hint: Consider the higher derivatives.]

   Give a generalization of this result and prove it.

(viii) A function $f : \mathbb{C} \to \mathbb{C}$ is called a **doubly periodic function** if there exists a constant $m \in \mathbb{R}$ such that

$$f(z + m) = f(z) \text{ and } f(z + mi) = f(z) \text{ for all } z \in \mathbb{C}.$$

Show that the only complex analytic doubly periodic functions and constant. [Hint: Integrate round a suitable square and show that f is constant on $\mathbb{C}$.]

(ix) Let $f : \mathbb{C} \to \mathbb{C}$ be complex differentiable and suppose that there exists a positive real constant M such that $|f(z)| \geq M$ for all $z \in \mathbb{C}$. Prove that f is constant.

(x) Let $f(z) = \exp(z^3)$. Find the maximum of $|f|$ on the set $D = \{z : |z| \leq 2\}$.

(xi) Let u be an harmonic function on an open set $D \subset \mathbb{C}$. Let γ be a closed simple contour with its image and interior in D. Show that u has no maximum on the interior of γ or is constant. [Hint: Consider the function $F(x + iy) = e^{u(x,y)+iv(x,y)}$ where v is the harmonic conjugate of u.]

(xii) Prove the **Minimum Modulus Principle**: Let $f : D \to \mathbb{C}$ be a differentiable complex function such that $f(z) \neq 0$ for all $z \in D$ and γ be a closed contour such that $\gamma^* \cup \text{Int}(\gamma) \subset D$. If $|f(z)| \geq M$ for all $z \in \gamma^*$, then $|f(w)| \geq M$ for all $w \in \text{Int}(\gamma)$.

Give a counterexample to show that the condition $f(z) \neq 0$ cannot be dropped from this statement.

(xiii) Let $f : D \to D$ be an analytic function where $D = \{z \in \mathbb{C} : |z| < 1\}$. Prove that $|f'(0)| \leq 1$. (Note that this says that the $f(0) = 0$ condition in Schwartz's Lemma is unnecessary to conclude $|f'(0)| \leq 1$.)

Summary

❏ Cauchy's theorem can be used to prove surprising theorems that have no analogues in real analysis.

❏ Fudamental Theorem of Algebra: Every complex polynomial has a complex root.

❏ Liouville's Theorem: Any differential function bounded on the whole of $\mathbb{C}$ is contant.

❏ The Maximum Modulus Principle: The modulus of a function on a domain achieves its maximum on the boundary of the domain.

CHAPTER 15

Laurent Series

In this chapter we move from differentiable functions to functions that are non-differentiable at isolated points. By introducing the notions of Laurent expansions and poles we will develop a very practical theory culminating in Cauchy's Residue Theorem which provides a simple method for calculating complicated integrals.

A motivating example

Let us consider an example: Integrate $\int_{C_1} \frac{e^{3z}}{z^2}\, dz$, where C_1 denotes the standard contour of a unit circle around the origin.

We can't use Cauchy's Integral Formula as the integrand is not linear in the denominator and the Fundamental Theorem of Calculus doesn't help as we can't see any obvious antiderivative. (In fact, no such antiderivative exists.)

Let's expand the integrand as a series, integrate term-by-term (assuming we

152

can) and let's see what happens:

$$\int_{C_1} \frac{e^{3z}}{z^2}\, dz = \int_{C_1} \frac{1}{z^2}\left(1 + 3z + \frac{(3z)^2}{2!} + \frac{(3z)^3}{3!} + \dots\right)\, dz$$

$$= \int_{C_1}\left(\frac{1}{z^2} + \frac{3}{z} + \frac{9}{2} + \frac{9z}{2} + \dots\right)\, dz$$

$$= \int_{C_1}\frac{1}{z^2}\, dz + \int_{C_1}\frac{3}{z}\, dz + \int_{C_1}\frac{9}{2}\, dz + \int_{C_1}\frac{9z}{2}\, dz + \dots$$

$$= 0 + 3 \times 2\pi i + 0 + 0 + \dots$$

$$= 6\pi i.$$

Notice that

$$\text{only the coefficient of the } \frac{1}{z} \text{ term matters.}$$

By the Fundamental Example all the other terms are irrelevant. This observation is key to understanding the subsequent theory. First we define a nice collection of non-differentiable functions.

Laurent series

Recall that an *analytic* function can be represented as

$$f(z) = \sum_{n=0}^{\infty} a_n(z - z_0)^n \text{ for } |z - z_0| < R.$$

All the powers of $z - z_0$ are positive. The integrand e^{3z}/z^2 in the example above can be represented by a series with some *negative* powers so we shall make a definition to cover this situation.

Definition 15.1
Suppose that $f : D \to \mathbb{C}$ is a complex function. Then we say f has a **Laurent expansion at** z_0 if there exist $a_n \in \mathbb{C}$, and real numbers R_1 and R_2 with $0 \leq R_1 < R_2$, such that

$$f(z) = \sum_{n=-\infty}^{\infty} a_n(z - z_0)^n$$

for all z with $R_1 < |z - z_0| < R_2$.

The series expression is called a **Laurent series**. The numbers R_1 and R_2 are called the **radii of the expansion**.

Remarks 15.2

(i) Note that the summation limits in the definition go from $-\infty$ to ∞, and that the expansion, does not necessarily equal $f(z_0)$ when $z = z_0$, (which may be undefinable anyway).

(ii) In this situation, convergence of $\displaystyle\sum_{n=-\infty}^{\infty} a_n(z - z_0)^n$ at z means that

$$\sum_{n=-\infty}^{-1} a_n(z - z_0)^n \qquad \text{and} \qquad \sum_{n=0}^{\infty} a_n(z - z_0)^n$$

converge.

(iii) A set of the form $\{z \in \mathbb{C} : R_1 < |z - z_0| < R_2\}$ is called an **annulus**. It is the area between the circles in the next diagram.

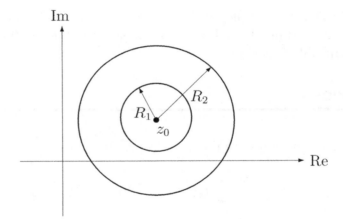

Examples 15.3

(i) The function $f(z) = \dfrac{1}{z - z_0}$ has a Laurent expansion at z_0 with only one non-zero term: $a_{-1} = 1$, $a_n = 0$ for $n \neq -1$, $R_1 = 0$ and $R_2 = \infty$.

(ii) In the example above, the expansion of $f(z) = \dfrac{e^{3z}}{z^2}$ at 0 is

$$\frac{e^{3z}}{z^2} = \frac{1}{z^2} + \frac{3}{z} + \frac{9}{2} + \frac{9z}{2} + \dots$$

This has $a_n = \dfrac{3^{n+2}}{(n + 2)!}$ for $n \geq 0$, $a_{-1} = 3$ and $a_{-2} = 1$, and $a_j = 0$ for $j \leq -3$.

To work out the radii of expansion we can calculate via the Ratio Test that the series defined by the positive exponents converges for all z. The sum of the two negative exponents is only undefined at the point $z = 0$. Hence, the radii of the expansion are $R_1 = 0$ and $R_2 = \infty$.

(iii) Any power series with a positive radius of convergence is a Laurent expansion. Hence, any function differentiable at z_0 has a Laurent expansion by Taylor's Theorem. In this case $a_n = 0$ for all $n < 0$.

In the above examples we see that $R_1 = 0$. This case is common enough to be given a definition.

Definition 15.4
The set $\{z \in \mathbb{C} \text{ such that } 0 < |z - z_0| < R \text{ for some } R > 0\}$ is called a **punctured neigbourhood** of z_0.

Remark 15.5
To save ourselves the hassle of specifying the R and writing $0 < |z - z_0| < R$, we often say that f has a **Laurent expansion near** z_0 or **in a neighbourhood of** z_0.

Examples 15.6
(i) We can calculate a Laurent expansion of $\dfrac{1}{\sin z}$ up to degree 3 near 0 as follows:

$$\frac{1}{\sin z} = \frac{1}{\left(z - \dfrac{z^3}{3!} + \dfrac{z^5}{5!} - \cdots\right)}$$

$$= \frac{1}{z\left(1 - \dfrac{z^2}{3!} + \dfrac{z^4}{5!} - \cdots\right)}$$

$$= \frac{1}{z}\left(\frac{1}{1 - \left(\dfrac{z^2}{3!} - \dfrac{z^4}{5!} + \cdots\right)}\right)$$

Now, let $g(z) = \dfrac{z^2}{3!} - \dfrac{z^4}{5!} + \dots$. This is a convergent series and hence is continuous near 0. Thus for small z we have $|g(z)| < 1$ and using $\dfrac{1}{1-x} =$

$\sum x^n$ we find

$$\frac{1}{\sin z} = \frac{1}{z}\left(1 + g(z) + g(z)^2 + \ldots\right)$$

$$= \frac{1}{z}\left(1 + \left(\frac{z^2}{3!} - \frac{z^4}{5!} + \ldots\right) + \left(\frac{z^2}{3!} - \frac{z^4}{5!} + \ldots\right)^2 + \ldots\right)$$

$$= \frac{1}{z}\left(1 + \frac{z^2}{3!} - \frac{z^4}{5!} + \frac{z^4}{(3!)^2} - \ldots\right)$$

$$= \frac{1}{z}\left(1 + \frac{z^2}{6} + \frac{7z^4}{360} + \ldots\right)$$

$$= \frac{1}{z} + \frac{z}{6} + \frac{7z^3}{360} + \ldots$$

We can take $R_1 = 0$ and note that for R_2 we are required to determine the z for which $|g(z)| < 1$. This is possible but rather time-consuming. Fortunately, in practice we often just need to know that some non-zero R_2 exists, i.e., some punctured neighbourhood exists; the precise size of the neighbourhood is unimportant.

(ii) From the previous example we can find a Laurent expansion for $\cot z$ valid in a punctured neighbourhood of 0 and up to degree 3.

Since $\cot z = \dfrac{\cos z}{\sin z}$ by definition we have

$$\cot z = \frac{\cos z}{\sin z}$$

$$= \left(1 - \frac{z^2}{2} + \frac{z^4}{4!} + \ldots\right)\left(\frac{1}{z} + \frac{z}{6} + \frac{7z^3}{360} + \ldots\right)$$

$$= \frac{1}{z} + \frac{z}{6} + \frac{7z^3}{360} + \ldots - \frac{z}{2} - \frac{z^3}{12} + \ldots + \frac{z^3}{24} + \ldots$$

$$= \frac{1}{z} - \frac{z}{3} - \frac{z^3}{45} + \ldots$$

Now for an example where the set of convergence is not a punctured neighbourhood.

Example 15.7

By expanding about different points we can produce different expansions for the same function. Let

$$f(z) = \frac{1}{z^2 - 5z + 6}.$$

Then

$$f(z) = \frac{1}{z-3} - \frac{1}{z-2}.$$

For $|z| < 2$ this function is analytic and hence can be written as a Taylor series. We have

$$\frac{1}{z-3} = -\frac{1}{3(1-(z/3))} = -\frac{1}{3}\sum_{n=0}^{\infty}\left(\frac{z}{3}\right)^n = -\sum_{n=0}^{\infty}\frac{z^n}{3^{n+1}}$$

for $|z/3| < 1$, i.e., $|z| < 3$. We also have

$$\frac{1}{z-2} = -\frac{1}{2(1-(z/2))} = -\frac{1}{2}\sum_{n=0}^{\infty}\left(\frac{z}{2}\right)^n = -\sum_{n=0}^{\infty}\frac{z^n}{2^{n+1}}$$

for $|z/2| < 1$, i.e., $|z| < 2$. Therefore, for $|z| < 2$, we have

$$\frac{1}{z^2-5z+6} = \frac{1}{z-3} - \frac{1}{z-2}$$

$$= -\sum_{n=0}^{\infty}\frac{z^n}{3^{n+1}} + \sum_{n=0}^{\infty}\frac{z^n}{2^{n+1}}$$

$$= \sum_{n=0}^{\infty}\left(\frac{1}{2^{n+1}} - \frac{1}{3^{n+1}}\right)z^n.$$

We can expand the function on the annulus $2 < |z| < 3$. We have

$$\frac{1}{z-2} = \frac{1}{z(1-(2/z))}$$

$$= \frac{1}{z}\sum_{n=0}^{\infty}\left(\frac{2}{z}\right)^n, \text{ for } |2/z| < 1, \text{ i.e., } 2 < |z|,$$

$$= \sum_{n=-\infty}^{0}\frac{z^{n+1}}{2^n}$$

$$= \sum_{n=-\infty}^{-1}\frac{z^n}{2^{n+1}}.$$

Therefore, for $2 < |z| < 3$, we have

$$\frac{1}{z^2 - 5z + 6} = \frac{1}{z - 3} - \frac{1}{z - 2}$$

$$= -\sum_{n=0}^{\infty} \frac{z^n}{3^{n+1}} - \sum_{n=-\infty}^{-1} \frac{z^n}{2^{n+1}}$$

$$= -\left(\sum_{n=-\infty}^{-1} \frac{z^n}{2^{n+1}} + \sum_{n=0}^{\infty} \frac{z^n}{3^{n+1}} \right).$$

We can also take the expansion about $z = 3$. First we expand $1/(z - 2)$ about $z = 3$.

$$\frac{1}{z - 2} = \frac{1}{(z - 3) + 1}$$

$$= \frac{1}{1 - (3 - z)}$$

$$= \sum_{n=0}^{\infty} (3 - z)^n, \text{ for } |3 - z| < 1,$$

$$= \sum_{n=0}^{\infty} (-1)^n (z - 3)^n.$$

Therefore, on $0 < |z - 3| < 1$, we have

$$\frac{1}{z - 3} - \frac{1}{z - 2} = \frac{1}{z - 3} - \sum_{n=0}^{\infty} (-1)^n (z - 3)^n$$

$$= -\sum_{n=-1}^{\infty} (-1)^n (z - 3)^n.$$

Exercise 15.8
Find expansions for the above function for $|z| > 3$ and $0 < |z - 2| < 1$.

Laurent's Theorem

We conclude the chapter with a single theorem: A function analytic within an annulus has a Laurent expansion on the annulus. It is useful to know but not, except for the uniqueness of coefficients, used in any subsequent statements or proofs other than exercises.

Theorem 15.9 (Laurent's Theorem)
Suppose that f is analytic on the annulus $R_1 < |z - z_0| < R_2$ then there exists a Laurent expansion of the form

$$f(z) = \sum_{n=-\infty}^{\infty} a_n(z - z_0)^n \qquad \text{on } R_1 < |z - z_0| < R_2$$

where the a_n are unique and are calculated by

$$a_n = \frac{1}{2\pi i} \int_{C_r} \frac{f(\zeta)}{(\zeta - z_0)^{n+1}} d\zeta$$

for any r with $R_1 < r < R_2$ and C_r is the circle contour of radius r centred at z_0.

First, let's show that the value of the integral does not depend on r.

Lemma 15.10
Suppose that f is analytic on the annulus $R_1 < |z - z_0| < R_2$. Then the number

$$\frac{1}{2\pi i} \int_{C_r} \frac{f(\zeta)}{(\zeta - z_0)^{n+1}} d\zeta$$

is independent of r for all $R_1 < r < R_2$.

Proof. Without loss of generality we can assume that $z_0 = 0$. We must show that if r_1 and r_2 are such that $R_1 < r_1 < R_2$ and $R_1 < r_2 < R_2$ then

$$\int_{C_{r_1}} \frac{f(\zeta)}{\zeta^{n+1}} d\zeta = \int_{C_{r_2}} \frac{f(\zeta)}{\zeta^{n+1}} d\zeta.$$

Again without loss of generality we can assume $r_1 < r_2$. Consider the contour $C_{r_1} + \beta - C_{r_2} - \beta$ as in the following diagram.

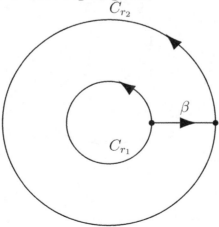

Since f is analytic on the annulus, by Cauchy's Theorem, we see

$$\int_{C_{r_1}+\beta-C_{r_2}-\beta} \frac{f(\zeta)}{\zeta^{n+1}} d\zeta = 0$$

$$\left(\int_{C_{r_1}} + \int_{\beta} + \int_{-C_{r_2}} + \int_{-\beta}\right)\left(\frac{f(\zeta)}{\zeta^{n+1}} d\zeta\right) = 0$$

$$\int_{C_{r_1}} \frac{f(\zeta)}{\zeta^{n+1}} d\zeta - \int_{C_{r_2}} \frac{f(\zeta)}{\zeta^{n+1}} = 0,$$

as the contributions from the β's cancel and $\int_{-C_{r_2}} = -\int_{C_{r_2}}$. $\qquad\qquad\square$

Proof (of Theorem).   [The proof works through a clever choice of contours and by using an argument similar to the one in the Taylor series proof.]

Without loss of generality we can assume $z_0 = 0$. Let z be any point in the annulus. Then there exists r and R such that $0 \leq R_1 < r < |z| < R < R_2$. Consider the contours C_r and C_R with two cross cuts.

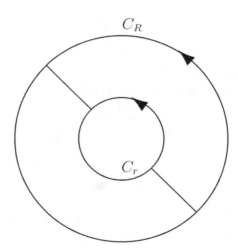

We can consider two contours γ and δ in the diagram where γ is the contour containing z:

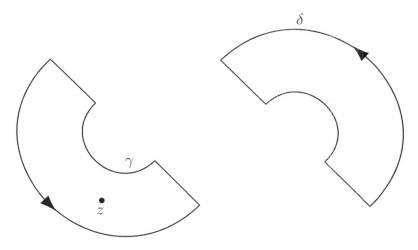

Then

$$\frac{1}{2\pi i}\int_{\gamma}\frac{f(\zeta)}{\zeta - z}d\zeta = f(z)$$

by Cauchy's Integral Formula, and

$$\frac{1}{2\pi i}\int_{\delta}\frac{f(\zeta)}{\zeta - z}d\zeta = 0$$

by Cauchy's Theorem. Thus their sum is equal to $f(z)$. However their sum is also equal to the sum of the integrals:

$$\frac{1}{2\pi i}\int_{C_R}\frac{f(\zeta)}{\zeta - z}d\zeta + \frac{1}{2\pi i}\int_{-C_r}\frac{f(\zeta)}{\zeta - z}d\zeta.$$

This is because the integrals from the cross cuts will cancel. Now we apply an argument similar to the one in the proof of the Taylor series theorem (Theorem 13.3).

For $\zeta \in C_R^*$ we have $\left|\dfrac{z}{\zeta}\right| < 1$ and hence

$$\frac{1}{\zeta - z} = \frac{1}{\zeta\left(1 - \dfrac{z}{\zeta}\right)} = \frac{1}{\zeta}\sum_{n=0}^{\infty}\frac{z^n}{\zeta^n} = \sum_{n=0}^{\infty}\frac{z^n}{\zeta^{n+1}}.$$

Hence

$$\frac{1}{2\pi i} \int_{C_R} \frac{f(\zeta)}{\zeta - z} d\zeta = \frac{1}{2\pi i} \int_{C_R} \sum_{n=0}^{\infty} \frac{f(\zeta)}{\zeta^{n+1}} z^n d\zeta$$

$$= \frac{1}{2\pi i} \sum_{n=0}^{\infty} \int_{C_R} \frac{f(\zeta)}{\zeta^{n+1}} z^n d\zeta, \text{ by same argument as in Theorem 13.3,}$$

$$= \sum_{n=0}^{\infty} \left(\frac{1}{2\pi i} \int_{C_R} \frac{f(\zeta)}{\zeta^{n+1}} d\zeta \right) z^n$$

$$= \sum_{n=0}^{\infty} a_n z^n, \text{ by Lemma 15.10.}$$

For $\zeta \in C_r^*$ we have $\left| \dfrac{\zeta}{z} \right| < 1$ and hence

$$\frac{1}{\zeta - z} = -\frac{1}{z\left(1 - \dfrac{\zeta}{z}\right)} = -\frac{1}{z} \sum_{n=0}^{\infty} \left(\frac{\zeta}{z} \right)^n = -\sum_{n=0}^{\infty} \frac{\zeta^n}{z^{n+1}}.$$

Therefore

$$\frac{1}{2\pi i} \int_{-C_r} \frac{f(\zeta)}{\zeta - z} d\zeta = -\frac{1}{2\pi i} \int_{C_r} \left(-\sum_{n=0}^{\infty} \frac{f(\zeta)\zeta^n}{z^{n+1}} \right) d\zeta$$

$$= \frac{1}{2\pi i} \sum_{n=0}^{\infty} \int_{C_r} \frac{f(\zeta)\zeta^n}{z^{n+1}} d\zeta$$

$$= \sum_{n=0}^{\infty} \left(\frac{1}{2\pi i} \int_{C_r} \frac{f(\zeta)}{\zeta^{-n}} d\zeta \right) z^{-(n+1)}$$

$$= \sum_{n=0}^{\infty} a_{-(n+1)} z^{-(n+1)}$$

$$= \sum_{n=-\infty}^{-1} a_n z^n.$$

All that remains is to show uniqueness. Suppose that there exists b_n such that

$f(z) = \displaystyle\sum_{n=-\infty}^{\infty} b_n z^n$ for $R_1 < |z| < R_2$. Then,

$$a_n = \frac{1}{2\pi i} \int_{C_r} \frac{f(\zeta)}{\zeta^{n+1}} d\zeta$$

$$= \frac{1}{2\pi i} \int_{C_r} \sum_{m=-\infty}^{\infty} \frac{b_m \zeta^m}{\zeta^{n+1}} d\zeta$$

$$= \frac{1}{2\pi i} \sum_{m=-\infty}^{\infty} \int_{C_r} b_m \zeta^{m-n-1} d\zeta, \text{ using the Weierstrass M-test,}$$

$$= \frac{1}{2\pi i} \sum_{m=-\infty}^{\infty} b_m \int_{C_r} \zeta^{m-n-1} d\zeta.$$

But by the fundamental example the integral is $2\pi i$ for $m - n - 1 = -1$, i.e. $m = n$ and 0 otherwise. Thus

$$a_n = \frac{1}{2\pi i} \left(\cdots + 0 + \cdots + 0 + 2\pi i\, b_n + 0 + \cdots + 0 + \ldots \right) = b_n.$$

$\square$

Exercises

Exercises 15.11

(i) Find a Laurent expansions for the following

(a) $\dfrac{\sin z}{z^4}$ defined on $\mathbb{C}\backslash\{0\}$,

(b) $\dfrac{e^{2z}}{(z-1)^3}$ about $z = 1$ and up to linear terms,

(c) $\dfrac{1}{z^2 - 1}$ about $z = 1$, (Hint: find the Taylor series of $1/(z+1)$ about $z = 1$ first),

(d) $\dfrac{1}{z^2 + 1}$ about $z = i$,

(e) $\dfrac{z}{(z+1)(z+2)}$ about $z = -2$.

(f) $\dfrac{z^3}{(2z+1)(3z-2)}$ at 0

(g) $\dfrac{1}{z^2(z^2+4)}$ at $z=0$ and $z=2$ up to degree 4, stating where these are valid,

(ii) Calculate the Laurent series at 0 for $\dfrac{(e^z-1)^2}{z^2}$. What do you notice?

(iii) Suppose that f is analytic and bounded by M in $r < |z - z_0| < R$. For the Laurent series $\sum_{n=-\infty}^{\infty} a_n(z-z_0)^n$ show that

$$|a_n| \le \frac{M}{R^n} \quad \text{and} \quad |a_{-n}| \le Mr^n$$

for $n = 0, 1, 2, 3, \ldots$.

(iv) Define the **nth Bessel function**, J_n, for $n \in \mathbb{Z}$, to be

$$J_n(w) = \frac{1}{2\pi} \int_{-\pi}^{\pi} \cos(n\theta - w\sin\theta)\, d\theta.$$

Using the contour $\gamma(t) = e^{-i\theta}$, $-\pi \le \theta \le \pi$, show that

$$\exp\left(\frac{w(z - z^{-1})}{2}\right) = \sum_{n=-\infty}^{\infty} J_n(w) z^n.$$

(v) Suppose that $h(z) = f(z) + g(z)$ on the annulus $r < |z| < R$ where f is analytic for $|z| < R$ and g is analytic and bounded for $|z| > r$. Let a_n be the Laurent coefficients for h in the annulus. Show that

$$g(z) = \alpha + \sum_{n=1}^{\infty} a_n z^n$$

for some $\alpha \in \mathbb{C}$ and all $|z| < R$.

Summary

❏ An expansion of f of the form

$$f(z) = \sum_{n=-\infty}^{\infty} a_n(z - z_0)^n$$

for all z with $R_1 < |z - z_0| < R_2$, is called a Laurent expansion at z_0.

❑ Suppose that f is analytic on the annulus $R_1 < |z - z_0| < R_2$ then there exists a Laurent expansion of the form

$$f(z) = \sum_{n=-\infty}^{\infty} a_n(z - z_0)^n \qquad \text{on } R_1 < |z - z_0| < R_2$$

where the a_n are unique and are calculated by

$$a_n = \frac{1}{2\pi i} \int_{C_r} \frac{f(\zeta)}{(\zeta - z_0)^{n+1}} d\zeta$$

for any r with $R_1 < r < R_2$ and C_r is the circle contour of radius r centred at z_0.

CHAPTER 16

Singularities, Zeros and Poles

At the start of the previous chapter we saw how the calculation of an integral involved analysis of the z^{-1} coefficient of a Laurent expansion. In this chapter we lay the groundwork for a thorough study of this coefficient in the next chapter.

Definition 16.1
Suppose that f is a complex function. We say that f has an **isolated singularity** at z_0 if there exists R such that f is analytic for all z such that $0 < |z - z_0| < R$ but not differentiable at z_0.

Examples 16.2

(i) Let $f : \mathbb{C} \to \mathbb{C}$ be defined by

$$f(z) = \begin{cases} z, & \text{for } z \neq i, \\ 1, & \text{for } z = i. \end{cases}$$

This is not complex differentiable at i since it is not even continuous there. Hence, f has an isolated singularity at i.

(ii) The function $f(z) = (\sin 2z)/z$ is analytic on $\mathbb{C}\backslash\{0\}$ but not defined at $z = 0$ and hence cannot be complex differentiable there. Thus, f has an isolated singularity at 0.

(iii) Let $f : \mathbb{C}\backslash\{0\} \to \mathbb{C}$ be given by $f(z) = z^2 + 3$. Then f is differentiable for $z \neq 0$ but is not differentiable at 0 since it is not defined there. Therefore, f has an isolated singularity at 0. Note that this is because of the choice of domain. The fact that we *can* define a function for all z will be dealt with soon.

(iv) Define f by

$$f(z) = \frac{1}{(z-1)(z-3)}.$$

Then, f has isolated singularities at $z = 1$ and $z = 3$ since f is differentiable everywhere except 1 and 3.

From Laurent's Theorem, Theorem 15.9, we know that at an isolated singularity there exists a Laurent expansion for f valid on $0 < |z - z_0| < R$ for some R. Using this expansion we can classify singularities into three types. Suppose that $f(z) = \sum\limits_{n=-\infty}^{\infty} a_n(z - z_0)^n$. Three possibilities can occur:

(i) All a_n are zero for $n < 0$.

(ii) A finite number of a_n are non-zero for $n < 0$.

(iii) An infinite number of a_n are non-zero for $n < 0$.

We shall give definitions for each of these situations and remark that it is the second one which will be the most important for us.

Removable and essential singularities

Let's deal with the first of the three possibilities.

Definition 16.3
Suppose that f has an isolated singularity and its Laurent expansion has $a_n = 0$ for all $n < 0$. Then we say that f has a **removable singularity**. This is because we can redefine f by changing its domain of definition to include z_0 if needed and we define $f(z_0) = a_0$

Examples 16.4
(i) The function in Examples 16.2(i),

$$f(z) = \begin{cases} z, & \text{for } z \neq i, \\ 1, & \text{for } z = i. \end{cases}$$

has Laurent expansion $f(z) = i + (z - i) = z$ for $0 < |z - i| < \infty$. Therefore, the singularity is removable. We can redefine f at i to be $f(i) = i$ to remove the singularity.

(ii) Recall that the function in Examples 16.2(ii), $f(z) = \dfrac{\sin 2z}{z}$ is analytic on $\mathbb{C}\backslash\{0\}$ but not defined at $z = 0$.

The Laurent expansion at 0 is

$$
\begin{aligned}
f(z) &= \frac{\sin 2z}{z} \\
&= \frac{1}{z}\left(2z - \frac{8z^3}{3!} + \frac{32z^5}{5!} - \cdots\right) \\
&= 2 - \frac{8z^2}{3!} + \frac{32z^4}{5!} - \cdots.
\end{aligned}
$$

The singularity is therefore removable. Since the limit of the right hand side as $z \to 0$ is 2 we can define $f(0) = 2$ to make f analytic on all of $\mathbb{C}$. That is,

$$
f(z) = \begin{cases} \dfrac{\sin 2z}{z}, & \text{for } z \neq 0, \\ 2, & \text{for } z = 0. \end{cases}
$$

(iii) The function in Examples 16.2(iii) has Laurent expansion $f(z) = z^2 + 3$ for $0 < |z - 0| < \infty$. By adding in the point $\{0\}$ to the domain of definition and simply setting $f(0) = 3$ we can remove the singularity at 0.

Remark 16.5
We can generalize the last set of examples. When f is not defined at the isolated singularity z_0 but has a removable singularity, then we can make f differentiable by defining $f(z_0)$ to be $\lim_{z \to z_0} f(z)$.

Next, before the second type of singularity, we consider the third type.

Definition 16.6
Suppose that f has an isolated singularity and its Laurent expansion has an infinite number of non-zero a_n for $n < 0$. Then we say f has an **essential singularity** at z_0.

Example 16.7
The function $f(z) = e^{1/z}$ has an essential singularity at 0 because

$$
f(z) = \sum_{n=0}^{\infty} \frac{(1/z)^n}{n!} = \sum_{n=0}^{\infty} \frac{z^{-n}}{n!} = \sum_{n=-\infty}^{0} \frac{z^n}{(-n)!},
$$

and this obviously has an infinite number of terms with negative exponent.

Near an essential singularity a function behaves wildly and since these are the hardest singularities to study we shall not look into them.

Zeros

We now wish to study our second possibility: a finite number of a_n are non-zero for $n < 0$. It turns out that these are closely related to zeros of functions.

Definition 16.8
A function $f : D \to \mathbb{C}$ has a **zero** at z_0 if $f(z_0) = 0$.

Examples 16.9
(i) The function sin has zeros at $k\pi$, for all $k \in \mathbb{Z}$. That these are the only zeros is an exercise. (The analogous statement and proof for cosine will be given in Examples 17.6.)

(ii) The function $f(z) = z^3(z^2 + 1)$ has a zero at $z = i$, one at $z = -i$, and one at $z = 0$.

(iii) By Theorem 3.1(iv) $f(z) = e^z$ has no zeros.

Exercise 16.10
Find the zeros of:

$$\text{(i) } z^2 + 9, \quad \text{(ii) } e^{z^2+9}(z + 4i), \quad \text{(iii) } e^{z-2} - 1.$$

We know that if $f : D \to \mathbb{C}$ is differentiable at z_0, it has a Taylor series expansion about z_0, i.e., $f(z) = \sum_{n=0}^{\infty} a_n(z - z_0)^n$ for all z with $|z - z_0| < R$ for some $R > 0$ and some $a_n \in \mathbb{C}$.

Definition 16.11
We say that an analytic function f has a **zero of order** m at z_0 if

$$a_0 = a_1 = \cdots = a_{m-1} = 0 \text{ but } a_m \neq 0.$$

This is also known as the **multiplicity** of f at z_0.

Remark 16.12
Obviously, by Theorem 13.3, f has a zero of order m at z_0 if and only if $f^{(j)}(z_0) = 0$ for all $j < m$ and $f^{(m)}(z_0) \neq 0$.

In particular, if $f(z_0) = 0$ and $f'(z_0) \neq 0$, then f has a zero of order 1. (By convention $f^{(0)}(z_0)$ is just $f(z_0)$.)

Examples 16.13
(i) The function $f(z) = z^2$ has a zero of order 2 at 0.

(ii) The function $f(z) = z(z + 2i)^3$ has a zero of order 1 at 0 and one of order 3 at $-2i$.

(iii) More generally, suppose that f is a polynomial with a root of multiplicity m at z_0. Then, f has a zero of order m at z_0. (Exercise!)

(iv) The function $f(z) = 2z^3 - (1 + 12i)z^2 + 6(i - 3)z + 9$ has a zero at $z = 3i$. Its multiplicity is 2:

$$\begin{aligned}
f'(z) &= 6z^2 - 2(1 + 12i)z + 6(i - 3) &\implies& \quad f'(3i) &=& \ 0, \\
f''(z) &= 12z - 2(1 + 12i) &\implies& \quad f''(3i) &=& \ -2 + 12i \neq 0.
\end{aligned}$$

(Since a polynomial of degree n has n roots when counted with multiplicity there must exist another root and it must have multiplicity 1.)

Exercise 16.14
Find the zeros and their orders of the following:

(i) $(z - 1)^3(z + 1)$ (ii) $(z^2 + 1)^2$,   (iii) ze^{z^2},   (iv) $(2z - 3i)^4$,   (v) $z^3 + 1$.

The proof of the following useful theorem is left as an exercise.

Theorem 16.15
Suppose that $f : D \to \mathbb{C}$ is an analytic function with a zero of order m at z_0. Then, there exists a differentiable function g and an $R > 0$, such that $f(z) = (z - z_0)^m g(z)$, for all $|z - z_0| < R$, and $g(z_0) \neq 0$.

This theorem will be used a number of times.

Poles

We now come to a central definition. The notion of pole, and that of residue to be introduced in the next chapter, are crucial in applying complex analysis to real problems. We have dealt with two of the three possibilities for the Laurent expansion for an isolated singularity. A pole is the third possibility: a finite number of the a_n are non-zero for $n < 0$.

Definition 16.16
Suppose that f has an isolated singularity at z_0. We say that the function has a **pole of order** N at z_0, where N is a positive integer, if there exists a Laurent expansion at z_0 of the form

$$f(z) = \sum_{n=-N}^{\infty} a_n(z - z_0)^n \quad (0 < |z - z_0| < R), \text{ with } a_{-N} \neq 0.$$

The order is also called the **multiplicity** of the pole. Poles of order $1, 2, 3 \ldots$ are called **simple, double, triple,** ...

The point is that for a pole there is only a *finite* number of negative exponents of $z - z_0$ in the expansion. For essential singularities, there is an infinite number of negative exponents.

Examples 16.17

(i) $\text{Order}(1/z^5) = 5$ at 0.

(ii) $\text{Order}(e^{3z}/z^2) = 2$ at 0 as $f(z) = \dfrac{1}{z^2} + \dfrac{3}{z} + \dfrac{9}{2} + \dfrac{9z}{2} + \ldots$.

(iii) $\text{Order}\left(\dfrac{i}{(z - 4i)^3} + (z - 4i)^2 - 2i(z - 4i)^5\right) = 3$ at $4i$.

(iv) $\text{Order}\left(\dfrac{e^z - 1}{z^3}\right) = 2$ at 0. We have

$$
\begin{aligned}
\frac{e^z - 1}{z^3} &= \frac{-1 + 1 + z + \dfrac{z^2}{2!} + \dfrac{z^3}{3!} + \ldots}{z^3} \\
&= \frac{1}{z^2} + \frac{1}{2z} + \frac{1}{6} + \ldots
\end{aligned}
$$

(v) From Example 15.7 we have on $0 < |z - 3| < 1$

$$
f(z) = \frac{1}{z^2 - 5z + 6} = -\sum_{n=-1}^{\infty} (-1)^n (z - 3)^n.
$$

Thus $\text{Order}(f) = 1$ at $z = 3$.

(vi) The function $f(z) = \dfrac{\cot z}{z^2}$ has a pole of order 3 at 0.

From Examples 15.6(ii) we know that in a neighbourhood of 0 and up to degree 3, that

$$
\cot z = \frac{1}{z} - \frac{z}{3} - \frac{z^3}{45} + \ldots
$$

Hence,

$$
\frac{\cot z}{z^2} = \frac{1}{z^3} - \frac{1}{3z} - \frac{z}{45} + \ldots
$$

which has a pole of order 3.

Although at a pole the function is not analytic there are some closely related functions that are analytic and these will prove useful later.

Theorem 16.18
Suppose f has a pole of order N at the point z_0 and that f is defined on $0 < |z - z_0| < R$. Prove that the following are true.

(i) The series $\sum_{n=0}^{\infty} a_n(z - z_0)^n$ converges for all z such that $|z - z_0| < R$ and hence is analytic at z_0.

(ii) The function $(z - z_0)^N f(z)$ is analytic at z_0.

The proof is an exercise.

How to find poles

You may have noticed that the order of a pole is often (but not always, see Examples 16.17(iv) and (vi) the same as the multiplicity of a zero in the denominator of a function's representation. The following makes this precise.

Lemma 16.19
Suppose that $f(z) = \dfrac{p(z)}{q(z)}$, where

(i) p is analytic at z_0, and $p(z_0) \neq 0$,

(ii) q is analytic at z_0, and has a zero of order N at z_0.

Then, f has a pole of order N at z_0.

Proof. By assumption and Theorem 16.15, on a neighbourhood of z_0, we have $q(z) = (z - z_0)^N r(z)$ where $r(z)$ is analytic and $r(z_0) \neq 0$. Then, $g(z) = \dfrac{p(z)}{r(z)}$ is analytic at z_0, so by Theorem 13.3 it has a power series expansion:

$$g(z) = \sum_{n=0}^{\infty} b_n(z - z_0)^n, \text{ valid for } |z - z_0| < R, \text{ some } R > 0.$$

Thus,

$$
\begin{aligned}
f(z) &= \frac{p(z)}{q(z)} \\
&= \frac{p(z)}{(z - z_0)^N r(z)} \\
&= \frac{g(z)}{(z - z_0)^N} \\
&= \frac{1}{(z - z_0)^N} \sum_{n=0}^{\infty} b_n (z - z_0)^n \\
&= \frac{b_0}{(z - z_0)^N} + \frac{b_1}{(z - z_0)^{N-1}} + \frac{b_2}{(z - z_0)^{N-2}} + \dots
\end{aligned}
$$

But,

$$
b_0 = g(z_0) = \frac{p(z_0)}{r(z_0)} \neq 0,
$$

because $p(z_0) \neq 0$ and $r(z_0) \neq 0$. So as $b_0 \neq 0$, we deduce that f has a pole of order N at z_0. $\qquad\square$

Common Error 16.20

The above lemma gives a method for finding poles and their orders. The definition of a pole is that given in Definition 16.16. It is not the multiplicity of the denominator in an expression. Do not confuse the *definition* of an object with a *process* by which we find that object.

Remark 16.21

Recall that a zero of order N of a polynomial is just a root of multiplicity N. This is useful for calculating the order of a pole.

Examples 16.22

(i) The function $\dfrac{\sin z}{(z - 3)^2}$ has a pole of order 2 at $z = 3$, since sin is analytic, $\sin 3 \neq 0$, and $(z - 3)^2$ is analytic with a root of multiplicity 2 at 3.

(ii) The quotient $\dfrac{z(z - 1)^4}{(z^2 - 2z + 5)^2}$ has poles of order 2 at $z = 1 \pm 2i$: Factorize the polynomial $z^2 - 2z + 5$ as $(z - (1 + 2i))(z - (1 - 2i))$. The resulting roots are repeated as

$$
(z^2 - 2z + 5)^2 = \left[(z - (1 + 2i))\right]^2 \left[(z - (1 - 2i))\right]^2.
$$

Exercises 16.23
Find the poles and their orders for

$$\text{(a) } \frac{1+e^z}{z(z^2-1)}, \quad \text{(b) } \frac{1+z}{z^3-2z^2}, \quad \text{(c) } \frac{e^{z^2}-1}{z^5} \text{ (Its order isn't 5!)}$$

Simple poles

As stated earlier poles of order 1 are called simple poles.

Examples 16.24
  (i) The function $1/z$ has a simple pole at 0.

  (ii) The function $\dfrac{\sin z}{4z^2}$ has a simple pole at 0:

$$\begin{aligned}
\frac{\sin z}{4z^2} &= \frac{z - \dfrac{z^3}{3!} + \dfrac{z^5}{5!} - \cdots}{4z^2} \\
&= \frac{1}{4z} - \frac{z}{3!4} + \frac{z^3}{5!4} - \cdots.
\end{aligned}$$

There is a useful condition for identifying simple poles.

Theorem 16.25
Suppose that $f(z) = \dfrac{p(z)}{q(z)}$ *where*

  (i) p *is analytic at* z_0, *and* $p(z_0) \neq 0$,

  (ii) q *is analytic at* z_0, $q(z_0) = 0$ *and* $q'(z_0) \neq 0$.

Then, f *has a simple pole at* z_0.

Proof. By Remark 16.12, q has a zero of order 1 at z_0. Thus, by Lemma 16.19, f has a pole of order 1 at z_0, i.e., the pole is simple. □

Examples 16.26
  (i) The function $f(z) = \dfrac{\cos z}{z}$ has a simple pole at 0: We have $p(z) = \cos z$ and $q(z) = z$, so $p(0) = 1 \neq 0$, $q(0) = 0$, and $q'(0) = 1 \neq 0$.

  (ii) The function $f(z) = \dfrac{z+3}{\sin z}$ has a simple pole at π: We have $p(z) = z+3$ and $q(z) = \sin z$, so $p(\pi) = \pi + 3 \neq 0$, $q(\pi) = 0$, and $q'(\pi) = \cos \pi = -1 \neq 0$.

Exercises

Exercises 16.27

(i) Find the zeros and their orders of the following functions

(a) $(z-2)^4 + (z-2)^3$,

(b) $\sin z$,

(c) $1 - \sin(z - \pi/2)$.

(ii) Show that if f is a polynomial with a root of multiplicity m at z_0, then f has a zero of order m at z_0.

(iii) Show that $\dfrac{3z}{1 - \cos z}$ has a pole of order 1 at 0.

(iv) Find and classify the singularities of the following.

(a) $\dfrac{2z^2 + 3z}{3z^2 - 11z - 4}$,

(b) $\dfrac{\sin(z + 3)}{z^6 - 9z^4}$,

(c) $\dfrac{1}{z \sin^2 z}$,

(d) $\tan^2 z$.

(e) $p(1/z)$ where p is a polynomial of degree m,

(f) $\cos\left(\dfrac{1}{z - \pi}\right)$.

(v) Suppose that $f(z) = p(z)/q(z)$ where p and q are analytic, p has a zero of order N_1 at z_0 and q has a zero of order $N_2 > N_1$ at z_0. Prove that f has a pole of order $N_2 - N_1$ at z_0.

Hence find the poles and their orders of the following functions:

(a) $f(z) = \dfrac{z + \pi}{\sin z}$,

(b) $f(z) = \dfrac{\sin(1 - \cos z)}{z^4}$,

(c) $f(z) = \dfrac{\sin^2 z}{(z - \pi)^3}$,

(d) $f(z) = \dfrac{z\cos(z) - \sin(z)}{z^5}$.

(vi) Prove Theorem 16.15.

(vii) Prove Theorem 16.18. Deduce from the theorem that the coefficients in such a Laurent expansion are unique. (That is we do not need to rely on Theorem 15.9.)

(viii) Suppose that f and g each have a pole at w. Prove or give a counterexample to the statement $\mathrm{Order}(fg) = \mathrm{Order}(f) + \mathrm{Order}(g)$ at the pole.

(ix) Replace the assumption of $p(z_0) \neq 0$ with $p(z_0) = 0$ in Lemma 16.25. Show that f has a removable singularity. Generalize this statement.

(x) Suppose that f and g are differentiable functions on the domain D and that $f(z_0) = 0$ for some $z_0 \in D$.

    (a) Show that $f(z) = (z - z_0)h(z)$, where h is differentiable for $|z - z_0| < R$ for some $R > 0$ and that $h(z_0) = f'(z_0)$.

    (b) Hence, deduce L'Hopital's Rule for complex differentiable functions: Show that, if $f(z_0) = g(z_0) = 0$ and $g'(z_0) \neq 0$, then

$$\lim_{z \to z_0} \frac{f(z)}{g(z)} = \frac{f'(z_0)}{g'(z_0)}.$$

(xi) Show that the following functions have removable singularities at the indicated points and give a value for the function at that point that makes the function differentiable there:

    (a) $f(z) = \dfrac{e^{z-3} - 1}{z - 3}$ at $z = 3$,

    (b) $f(z) = \dfrac{1 - \cos z}{z}$ at $z = 0$,

    (c) $f(z) = \dfrac{1 - \cos z}{z^2}$ at $z = 0$.

    (d) $f(z) = \cot(z) - 1/z$ at $z = 0$

(xii) Suppose that D is an open disc centred at z_0 and f is a differentiable function on $D \backslash \{z_0\}$. Prove that if f is bounded on $D \backslash \{z_0\}$, then f has a removable singularity.

(xiii) Prove that if f has an essential singularity at z, then f^p has an essential singularity at z for all $p \in \mathbb{N}$.

(xiv) Suppose that f has an isolated singularity at z_0 and that

$$|f(z)| \leq \frac{C}{|z - z_0|^{1+\epsilon}}, \text{ for } z \neq z_0,$$

where $\epsilon > 0$ and $C \in \mathbb{R}$. Show that f has removable singularity at z_0.

Summary

- A function $f : D \to \mathbb{C}$ has a zero at z_0 if $f(z_0) = 0$.

- Suppose $f(z) = \sum_{n=0}^{\infty} a_n(z - z_0)^n$ for all z with $|z - z_0| < R$ for some R. Then f has a zero of order m at z_0 if

$$a_0 = a_1 = \cdots = a_{m-1} = 0 \text{ but } a_m \neq 0.$$

- We say that f has a pole of order N at $z_0 \in D$, if there exists a Laurent expansion at z_0 of the form

$$f(z) = \sum_{n=-N}^{\infty} a_n(z - z_0)^n \ (0 < |z - z_0| < R), \text{ with } a_{-N} \neq 0.$$

- Suppose that $f(z) = \dfrac{p(z)}{q(z)}$, where

  (i) p is analytic at z_0, and $p(z_0) \neq 0$,

  (ii) q is analytic at z_0, and has a zero of order N at z_0.

  Then, f has a pole of order N at z_0.

- A simple pole is a pole of order 1.

CHAPTER 17

Residues

We saw in the introduction to Chapter 15 that to integrate $\dfrac{e^{3z}}{z^2}$ around the unit circle all we needed was the coefficient of the z^{-1} term of the Laurent expansion about zero. Let us now give that number an official name.

Definition 17.1
Suppose that f has a Laurent expansion in the neighbourhood of z_0. The **residue of f at** z_0, denoted $\mathrm{res}(f, z_0)$, is the coefficient a_{-1} in the expansion $\sum_{n=-\infty}^{\infty} a_n(z - z_0)^n$.

Examples 17.2
  (i) From Example 15.3(ii), $\mathrm{res}(e^{3z}/z^2, 0) = 3$.

 (ii) We have $\mathrm{res}(z^2, 0) = 0$. In fact, the residue is always 0 at differentiable points since differentiable functions have no negative exponents in their expansion by Taylor's Theorem.

(iii) From Examples 16.24(ii),

$$\frac{\sin z}{4z^2} = \frac{1}{4z} - \frac{z}{3!4} + \frac{z^3}{5!4} - \cdots$$

so $\mathrm{res}\left(\dfrac{\sin z}{4z^2}, 0\right) = \dfrac{1}{4}.$

Remark 17.3

The residue at a point is unique because the coefficients of a Laurent expansion are unique (see Theorem 15.9 or, for the case of a finite number of negative exponents, Exercise 16.27(vii)).

Bluffer's Guide to Calculating Residues

Residues are an important part of any Complex Analysis course. This section contains the main methods for calculating the residues of poles. You should know them and their proofs.

Method A: Simple Poles

Suppose that f has a *simple pole* (i.e., has order 1) at w. Then

$$\text{res}(f, w) = \lim_{z \to w} (z - w) f(z).$$

Proof. As f has a simple pole at w, then near w, for some $a_i \in \mathbb{C}$, we have

$$f(z) = \sum_{n=-1}^{\infty} a_n (z - w)^n.$$

So,

$$
\begin{aligned}
\lim_{z \to w} (z - w) f(z) &= \lim_{z \to w} (z - w) \sum_{n=-1}^{\infty} a_n (z - w)^n \\
&= \lim_{z \to w} (z - w) \left(\frac{a_{-1}}{z - w} + a_0 + a_1(z - w) + a_2(z - w)^2 + \dots \right) \\
&= \lim_{z \to w} a_{-1} + a_0(z - w) + a_1(z - w)^2 + a_2(z - w)^3 + \dots \\
&= a_{-1} \\
&= \text{res}(f, w).
\end{aligned}
$$

$\square$

Examples 17.4

(i) The function $f(z) = \dfrac{\sin(z)}{z - 3}$ has a pole of order 1 at $z = 3$, so

$$\text{res}(f, 3) = \lim_{z \to 3} (z - 3) f(z) = \lim_{z \to 3} (z - 3) \frac{\sin(z)}{z - 3} = \lim_{z \to 3} \sin(z) = \sin(3).$$

(ii) The function $f(z) = \dfrac{\sin(z)}{z^4(z-2)}$ has a pole of order 1 at $z = 2$, so

$$\text{res}(f, 2) = \lim_{z \to 2}(z-2)f(z) = \lim_{z \to 2}(z-2)\frac{\sin(z)}{z^4(z-2)} = \lim_{z \to 2}\frac{\sin(z)}{z^4} = \frac{\sin(2)}{16}.$$

(iii) More generally, suppose that f is analytic at w. Then $\text{res}\left(\dfrac{f(z)}{z-w}, w\right) = f(w)$:

$$\text{res}(f, w) = \lim_{z \to w}(z-w)\frac{f(z)}{z-w} = \lim_{z \to w}f(z) = f(w).$$

The two previous examples can be seen as examples of this.

(iv) The function $f(z) = \dfrac{3z}{1 - \cos z}$ has a pole of order 1 at $z = 0$ (see Exercise 16.27(iii)), so

$$\begin{aligned}
\text{res}(f, 0) &= \lim_{z \to 0}z\frac{3z}{1 - \cos z} = \lim_{z \to 0}\frac{3z^2}{2\sin^2(z/2)} = \lim_{z \to 0}\frac{12\,(z/2)^2}{2\sin^2(z/2)} \\
&= 6\lim_{z \to 0}\left(\frac{z/2}{\sin(z/2)}\right)^2 = 6\left(\lim_{z \to 0}\frac{z/2}{\sin(z/2)}\right)^2 = 6.
\end{aligned}$$

Method B: Simple poles for quotients (Really useful method!)

Suppose $f(z) = \dfrac{p(z)}{q(z)}$ with p and q analytic, $p(w) \neq 0$, $q(w) = 0$ and $q'(w) \neq 0$. Then,

$$\text{res}(f, w) = \frac{p(w)}{q'(w)}.$$

Proof. We have

$$\begin{aligned}
\frac{p(w)}{q'(w)} &= \frac{p(w)}{\lim_{z \to w}\dfrac{q(z) - q(w)}{z - w}}, \quad \text{by definition of differentiation} \\
&= \frac{\lim_{z \to w}p(z)}{\lim_{z \to w}\dfrac{q(z)}{z - w}} \quad \text{as } q(w) = 0 \text{ and } p \text{ is continuous,} \\
&= \lim_{z \to w}(z - w)\frac{p(z)}{q(z)} \\
&= \lim_{z \to w}(z - w)f(z) \\
&= \text{res}(f, w).
\end{aligned}$$

The last equality follows from Method A as, by Theorem 16.25, f has a simple pole at w.   □

Remark 17.5
The conditions imply that f has a simple pole, so the method will only work in this case.

Examples 17.6

(i) The function $f(z) = \dfrac{5}{1 - z^3}$ has a pole at $z = 1$. We have $p(z) = 5$ and $q(z) = 1 - z^3$ so $q'(1) = -3 \times 1^2 = -3$. Thus $\operatorname{res}(f, 1) = \dfrac{p(1)}{q'(1)} = \dfrac{5}{-3} = -\dfrac{5}{3}$.

(ii) Let $f(z) = \dfrac{2z^3 - 7}{z^2 - 4}$. Then by Lemma 16.25 there are simple poles at $z = \pm 2$.

Hence,
$$\operatorname{res}(f, 2) = \left. \frac{2z^3 - 7}{2z} \right|_{z=2} = \frac{9}{4},$$
and
$$\operatorname{res}(f, -2) = \left. \frac{2z^3 - 7}{2z} \right|_{z=-2} = \frac{23}{4}.$$

(iii) The function $f(z) = \dfrac{2z^2}{1 + z^4}$ has a pole at $z = e^{i\pi/4}$ (and more besides, but we'll ignore them). Let
$$p(z) = 2z^2 \text{ and } q(z) = 1 + z^4.$$
Then $p(e^{i\pi/4}) = 2e^{i\pi/2} \neq 0$; $q'(z) = 4z^3$ so $q'(e^{i\pi/4}) = 4e^{3i\pi/4} \neq 0$. Therefore,
$$\operatorname{res}\left(f, e^{i\pi/4}\right) = \frac{p(e^{i\pi/4})}{q'(e^{i\pi/4})} = \frac{2e^{i\pi/2}}{4e^{3i\pi/4}} = \frac{e^{-i\pi/4}}{2} = \frac{\sqrt{2}}{4}(1 - i).$$

(iv) Calculate the residues of the poles of $\tan z$.

Solution: First we find the poles. Since $\tan z = \dfrac{\sin z}{\cos z}$ and both sin and cos are analytic the poles are only possible at $\cos z = 0$. Let $z = x + iy$. Then
$$\begin{aligned} \cos z = 0 &\iff |\cos z|^2 = 0 \\ &\iff |\cos(x + iy)|^2 = 0 \\ &\iff \cos^2 x + \sinh^2 y = 0, \text{ by Exercise 3.16(xiii)}, \\ &\iff \cos^2 x = 0 \text{ and } \sinh^2 y = 0. \end{aligned}$$

Thus $x = \dfrac{\pi}{2} + k\pi$ for $k \in \mathbb{Z}$ and $y = 0$. Hence, the zeros lie only on the real line in the familiar places.

Since $\sin(\pi/2 + k\pi) \neq 0$ and

$$\frac{d}{dz}(\cos z)\bigg|_{z=\pi/2+k\pi} \neq 0$$

then by Theorem 16.25 the poles of $\tan z$ are simple. Hence, by Method B, for each $k \in \mathbb{Z}$, we have

$$\mathrm{res}(\tan z, \pi/2 + k\pi) = \lim_{z \to \pi/2+k\pi} \frac{\sin(\pi/2 + k\pi)}{-\sin(\pi/2 + k\pi)} = \lim_{z \to \pi/2+k\pi} -1 = -1.$$

(v) Suppose that

$$f(z) = \frac{z^3 + 2}{z^5 + 2iz^4 + 5z^3 - 2iz^2 - 2iz - 2}.$$

Calculate $\mathrm{res}(f, i)$.

Solution: Let $p(z) = z^3 + 2$ and $q(z) = z^5 + 2iz^4 + 5z^3 - 2iz^2 - 2iz - 2$. Then $p(i) = i^3 + 2 = -i + 2 \neq 0$ and $q'(z) = 5z^4 + 8iz^3 + 15z^2 - 4iz - 2i$ so $q'(i) = 2 - 2i \neq 0$. Thus f has a simple pole at $z = i$.

Then

$$\mathrm{res}(f, i) = \frac{-i + 2}{2 - 2i} = \frac{-i + 2}{2 - 2i}\frac{2 + 2i}{2 + 2i} = \frac{-2i + 2 + 4 + 4i}{4 + 4} = \frac{6 + 2i}{8} = \frac{3 + i}{4}.$$

Note that using Method A here requires factorising a quintic polynomial. This is not impossible but is best avoided!

Method C: Poles of order N

We now generalise Method A (which is the case $N = 1$ in the following).

Suppose that f has a pole of order N at w. Then

$$\mathrm{res}(f, w) = \frac{1}{(N-1)!} \lim_{z \to w} \frac{d^{N-1}}{dz^{N-1}} \left\{(z - w)^N f(z)\right\}.$$

Proof. By assumption $f(z) = \sum_{n=-N}^{\infty} a_n(z-w)^n$ for some $a_i \in \mathbb{C}$, $R \in \mathbb{R}$ and $0 < |z-w| < R$. Thus,

$$
\begin{aligned}
f(z) &= \sum_{n=-N}^{\infty} a_n(z-w)^n \\
(z-w)^N f(z) &= (z-w)^N \left(\frac{a_{-N}}{(z-w)^N} + \cdots + \frac{a_{-1}}{z-w} + a_0 + a_1(z-w) + \ldots \right) \\
&= a_{-N} + a_{-N+1}(z-w) + \cdots + a_{-1}(z-w)^{N-1} \\
&\quad + a_0(z-w)^N + \ldots.
\end{aligned}
$$

Therefore,

$$
\begin{aligned}
&\frac{1}{(N-1)!} \lim_{z \to w} \frac{d^{N-1}}{dz^{N-1}} \left\{ (z-w)^N f(z) \right\} \\
&= \frac{1}{(N-1)!} \lim_{z \to w} \frac{d^{N-1}}{dz^{N-1}} \left\{ a_{-N} + a_{-N+1}(z-w) + \ldots \right. \\
&\qquad \left. + a_{-1}(z-w)^{N-1} + a_0(z-w)^N + \ldots \right\} \\
&= \frac{1}{(N-1)!} \lim_{z \to w} \left\{ (N-1)! \, a_{-1} + N! \, a_0(z-w) + \ldots \right\} \\
&= \frac{1}{(N-1)!} (N-1)! \, a_{-1} \\
&= a_{-1} \\
&= \mathrm{res}(f, w).
\end{aligned}
$$

$\square$

Examples 17.7

(i) The function $f(z) = e^{\sin 3z}/z^2$ has a pole of order 2 at $z = 0$, (by Lemma 16.19). So,

$$
\begin{aligned}
\mathrm{res}(f, 0) &= \frac{1}{1!} \lim_{z \to 0} \frac{d}{dz} \left(z^2 \frac{e^{\sin 3z}}{z^2} \right) \\
&= \lim_{z \to 0} \frac{d}{dz} \left(e^{\sin 3z} \right) \\
&= \lim_{z \to 0} 3 \cos(3z) e^{\sin 3z} \\
&= 3 \times 1 \times e^0 \\
&= 3.
\end{aligned}
$$

(ii) Let $f(z) = \dfrac{1}{(z^2 + 16)^3}$.   As $z^2 + 16 = (z + 4i)(z - 4i)$ we see, again by Lemma 16.19, that this has poles of order 3 at $z = \pm 4i$. We can calculate the residues for both at the same time.

We have

$$
\begin{aligned}
\operatorname{res}(f, \pm 4i) &= \frac{1}{(3 - 1)!} \lim_{z \to \pm 4i} \frac{d^{3-1}}{dz^{3-1}} \left\{ (z \mp 4i)^3 \frac{1}{(z + 4i)^3 (z - 4i)^3} \right\} \\
&= \frac{1}{2!} \lim_{z \to \pm 4i} \frac{d^2}{dz^2} \frac{1}{(z \pm 4i)^3} \\
&= \frac{1}{2} \lim_{z \to \pm 4i} \frac{(-4)(-3)}{(z \pm 4i)^5} \\
&= \frac{6}{(\pm 4i \pm 4i)^5} \\
&= \pm \frac{6}{8^5 i^5} \\
&= \pm \frac{6}{32768 i} \\
&= \mp \frac{3}{16384} i.
\end{aligned}
$$

That is,

$$
\operatorname{res}(f, 4i) = -\frac{3}{16384} i \qquad \text{and} \qquad \operatorname{res}(f, -4i) = \frac{3}{16384} i.
$$

Method D: Direct expansion

This method is sometimes described as a last resort but as the second example shows it can be very efficient. In this method we expand functions as power series, etc., and then calculate coefficients. (This method gives good practice in manipulation of power series.)

Examples 17.8
  (i) Calculate $\operatorname{res}\left(\dfrac{\pi \cot \pi z}{z^2}, 0 \right)$.

From Examples 15.6(ii), we have

$$
\begin{aligned}
\frac{\pi \cot \pi z}{z^2} &= \frac{\pi}{z^2} \left(\frac{1}{\pi z} - \frac{\pi z}{3} - \frac{(\pi z)^3}{45} + \cdots \right) \\
&= \frac{1}{\pi z^3} - \frac{\pi}{3z} - \frac{\pi^3}{45} + \cdots
\end{aligned}
$$

Hence,

$$\text{res}\left(\frac{\pi \cot \pi z}{z^2}, 0\right) = -\frac{\pi}{3}.$$

Perhaps surprisingly, this example will be used in Chapter 21 to determine the sum of $\sum_{n=1}^{\infty} 1/n^2$. (Note that this is another good example where the order of the pole is not equal to the degree of the denominator.)

(ii) Calculate $\text{res}\left(\frac{\sin(z^2)}{z^{151}}, 0\right)$.

We have

$$\frac{\sin(z^2)}{z^{151}} = \frac{1}{z^{151}}\left(z^2 - \frac{(z^2)^3}{3!} + \frac{(z^2)^5}{5!} - \cdots\right)$$

$$= \frac{1}{z^{151}}\left(z^2 - \frac{z^6}{3!} + \frac{z^{10}}{5!} - \cdots - \frac{z^{150}}{75!} + \cdots\right).$$

Thus $\text{res}\left(\frac{\sin(z^2)}{z^{151}}, 0\right) = -\frac{1}{75!}.$

Note that the first term of the Laurent expansion at 0 is z^2/z^{151}, i.e., the pole is of order 149. Consequently, applying Method C would be time consuming!

(iii) Let $f(z) = \dfrac{z \sin(3\pi z)}{(z-1)^3}$. This has a pole of order 3 at $z = 1$. So, expand about $w = 1$: (i.e., let $z = w + h$ and take h as our variable),

$$f(1+h) = \frac{(1+h)\sin(3\pi(1+h))}{(1+h-1)^3}$$

$$= \frac{(1+h)\sin(3\pi + 3\pi h)}{h^3}$$

$$= -\frac{(1+h)\sin(3\pi h)}{h^3}$$

$$= -\frac{(1+h)}{h^3}\left(3\pi h - \frac{9}{2}\pi^3 h^3 + \frac{81}{40}\pi^5 h^5 + \cdots\right)$$

$$= -\frac{1}{h^3}\left(3\pi h + 3\pi h^2 - \frac{9}{2}\pi^3 h^3 - \frac{9}{2}\pi^3 h^4 + \frac{81}{40}\pi^5 h^5 + \cdots\right)$$

$$= -\frac{3\pi}{h^2} - \frac{3\pi}{h} + \frac{9}{2}\pi^3 + \frac{9}{2}\pi^3 h - \frac{81}{40}\pi^5 h^2 + \cdots$$

So $\text{res}(f, 1) = -3\pi$.

Note that in this case we calculated more terms than necessary as we went up to degree 2 rather than degree -1. Obviously it is better to have too many redundant terms than have the h^{-1} term wrong.

(iv) Find the residue of $f(z) = \dfrac{z^3 e^{iz}}{(z^2+1)^2}$ at $z = i$.

Solution: As before let us get an expansion by setting $z = i + h$. We have

$$f(i+h) = \frac{(i+h)^3 e^{i(i+h)}}{((i+h)^2+1)^2}$$

$$= \frac{(i+h)^3 e^{-1+ih}}{(-1+2ih+h^2+1)^2}$$

$$= \frac{(i+h)^3 e^{-1+ih}}{h^2(2i+h)^2}$$

$$= \frac{(i+h)^3 e^{-1+ih}}{h^2 \left(2i \left(1 - \dfrac{ih}{2} \right) \right)^2}$$

$$= \frac{(i+h)^3 e^{-1+ih}}{-4h^2 \left(1 - \dfrac{ih}{2} \right)^2}$$

$$= -\frac{(i+h)^3 e^{-1+ih}}{4h^2} \left(1 + \frac{ih}{2} + \left(\frac{ih}{2} \right)^2 + \dots \right)^2 .$$

We need only the h^{-1} term so rather than expand out everything in the above we look for certain terms. Since the denominator is h^2 we look for the coefficient of the h term in the numerator. To achieve this we need to look at all terms up to h. Thus we see

$$f(i+h) = -\frac{1}{4h^2} \left(-i - 3h + \dots \right) \frac{(1+ih+\dots)}{e} \left(1 + \frac{ih}{2} + \dots \right)^2$$

$$= \frac{1}{4eh^2} \left(i + 3h + \dots \right) \left(1 + ih + \dots \right) \left(1 + ih + \dots \right)$$

$$= \frac{1}{4eh^2} \left(i + 3h + \dots \right) \left(1 + 2ih + \dots \right)$$

$$= \frac{1}{4eh^2} \left(i + h + \dots \right)$$

Thus, the required residue is $\dfrac{1}{4e}$.

Obviously, this method can involve heavy calculations. (See Chapter 23 for a method for simplifying calculations.)

Method E: Double Poles

The previous four methods are suitable for most situations. The following method can be skipped in a first reading. It is included to show that other methods are possible. Here we describe a formula for the residue of a double pole, i.e., a pole of order 2.

Theorem 17.9
Suppose that $f(z) = \dfrac{p(z)}{q(z)}$ where p and q are analytic at w, $p(w) \neq 0$, $q(w) = q'(w) = 0$ and $q''(w) \neq 0$. Then f has a double pole at w and

$$\operatorname{res}(f, w) = \frac{2}{(q''(w))^2} \det \begin{vmatrix} p'(w) & p(w) \\ \frac{1}{3}q'''(w) & q''(w) \end{vmatrix}.$$

Proof. From the assumptions and Lemma 16.19 we see that q has a zero of order 2 and hence $q(z) = (z - w)^2 \, r(z)$ for some r analytic and non-zero at w. We have

$$\operatorname{res}\left(\frac{p}{q}, w\right) = \frac{1}{(2-1)!} \lim_{z \to w} \frac{d}{dz}\left((z-w)^2 \frac{p(z)}{q(z)}\right)$$

$$= \lim_{z \to w} \frac{d}{dz}\left(\frac{p(z)}{r(z)}\right)$$

$$= \lim_{z \to w} \frac{p'(z)r(z) - p(z)r'(z)}{(r(z))^2}$$

$$= \frac{1}{(r(w))^2} \det \begin{vmatrix} p'(w) & p(w) \\ r'(w) & r(w) \end{vmatrix}.$$

It is easy to calculate $q''(w) = 2r(w)$ and $q'''(w) = 6r'(w)$. Thus,

$$\operatorname{res}\left(\frac{p}{q}, w\right) = \frac{2^2 \cdot}{(q''(w))^2} \det \begin{vmatrix} p'(w) & p(w) \\ \frac{1}{2}\frac{1}{3}q'''(w) & \frac{1}{2}q''(w) \end{vmatrix}.$$

Taking out the 1/2 factor from the determinant gives the desired equality. □

Example 17.10
Calculate the residue $\operatorname{res}\left(\dfrac{3z^3 + 2z + 5}{z \sin z}, 0\right)$.

Solution: Let $p(z) = 4z^2 + 2z + 5$ and $q(z) = z \sin z$. Then we have

Function	Value at 0
$p(z) = 3z^3 + 2z + 5$	5
$p'(z) = 6z^2 + 2$	2
$q(z) = z \sin z$	0
$q'(z) = z \cos z + \sin z$	0
$q''(z) = -z \sin z + 2 \cos z$	2
$q'''(z) = -3 \sin z - z \cos z$	0

From Method E we get

$$\mathrm{res}\left(\frac{p}{q}, 0\right) = \frac{2}{2^2} \left| \begin{matrix} 2 & 5 \\ \frac{1}{3} \times 0 & 2 \end{matrix} \right| = \frac{1}{2} \times 4 = 2.$$

Note that we can use Method C but for this we need to calculate

$$\lim_{z \to 0} \frac{d}{dz} \left(\frac{3z^4 + 2z^2 + 5z}{\sin z} \right).$$

The derivative here is $\dfrac{g(z)}{\sin^2(z)}$ where $g(z)$ has 6 terms. Furthermore, to calculate the residue from this we really do have to calculate a limit, we can't simply substitute $z = 0$ since then $\sin^2 z = 0$. Although not complicated the calculation of this limit is tedious and therefore best avoided.

Corollary 17.11
Suppose that $f(z) = \dfrac{1}{q(z)}$ where $q(z)$ has a zero of order 2 at w. Then

$$\mathrm{res}(f, w) = -\frac{2}{3} \frac{q'''(w)}{q''(w)^2}$$

Proof. Follows immediately from the theorem.   □

Example 17.12
Find the residue of $1/(z^3 - 3z^2 + 4)$ at $z = 2$.
    Solution: Let $q(z) = z^3 - 3z^2 + 4$. Then we have

$$q'(z) = 3z^2 - 6z$$
$$q''(z) = 6z - 6$$
$$q'''(z) = 6.$$

Thus,

$$\mathrm{res}(1/q, 2) = -\frac{2}{3} \frac{6}{6^2} = -\frac{1}{9}.$$

Remark 17.13

Care needs to be taken with the choice of p and q in Methods such as B and E.
For example, the function $f(z) = \dfrac{\sin(z)}{z^3}$ has a pole of order 2 at $z = 0$. (To see this just calculate the Laurent expansion using the Taylor series for $\sin z$.)

We cannot use $p(z) = \sin z$ and $q(z) = z^3$ in Method E even though it is tempting to do so. This is because although p satisfies the conditions of the theorem q does not.

However, we can use $p(z) = \dfrac{\sin z}{z}$ and $q(z) = z^2$. In this case we can see that p is analytic at 0 if we remove the singularity and get $p(0) = 1$. Using the Taylor series for the removed singularity we can calculate $p'(0) = 0$. (If we tried to calculate it naively using the quotient rule then we get $\dfrac{z \cos z - \sin z}{z^2}$ and this is undefined at zero.)

Having said that, probably the best way to calculate $\dfrac{\sin z}{z^3}$ is to use Method D.

Exercises

Exercises 17.14

(i) Find the residues at all poles unless otherwise stated.

(a) $\dfrac{e^{iz}}{z^2 + 4}$ at $2i$,   (b) $\dfrac{e^{\sin z}}{z(z-2)^2}$ at 0,   (c) $\dfrac{e^{iz}}{z^3 - 1}$ at 1,

(d) $\dfrac{e^z}{z^6(z-1)}$,   (e) $\dfrac{\sin z}{z^6(z-1)}$,   (f) $\dfrac{\cos 2z}{z^3}$ at 0,

(g) $\dfrac{1}{e^z - 1}$ at 0,   (h) $\dfrac{1}{(e^z - 1)^2}$ at 0,   (i) $\dfrac{1 + \cos z}{(z - \pi)^3}$ at π,

(j) $\dfrac{e^{e^z}}{z^2}$,   (k) $\cot z$,   (l) $\dfrac{\cos(\pi e^{\pi z})}{z(z + 2i)^2}$,

(m) $\dfrac{\sin z}{(z - \pi)^6}$ at π,   (n) $\dfrac{\cos^2(2z^2)}{z^5}$ at 0.

(ii) Use Method E to solve the problem in Example 17.8(iv).

(iii) Find the poles for the function $\dfrac{1}{z^n - 1}$ and show that if ω is a pole, then the corresponding residue is $\dfrac{\omega}{n}$.

(iv) Suppose that p is an analytic function with $p(-\beta/\alpha) \neq 0$ where $\alpha, \beta \in \mathbb{C}$ and $\alpha \neq 0$. Show that

$$\text{res}\left(\frac{p(z)}{\alpha z + \beta}\right) = \frac{p(-\beta/\alpha)}{\alpha}.$$

(v) Suppose that $g : D \to \mathbb{C}$ is analytic function. Show that if $w \in D$ and $g(w) \neq 0$, then the function $f(z) = \dfrac{g(z)}{(z - w)^k}$ has a pole of order k at w and

$$\text{res}(f, w) = \frac{g^{(k-1)}(w)}{(k - 1)!}.$$

(vi) Suppose that $f : D \to \mathbb{C}$ is analytic on the open set D. Show that if f has a zero of order k at w, then $1/f$ has a pole of order k at w.

(vii) Suppose that f has a zero of order m at w. Show that $\text{res}(f'/f) = m$.

(viii) Find the poles and their residues for

$$\frac{1}{z^n + z^{n-1} + z^{n-2} + \cdots + 1}$$

where n is a positive integer.

(ix) We can calculate residues of double poles by a repeated limits method: Suppose that f has a double pole at w. Let $c = \lim_{z \to w}(z - w)^2 f(z)$. Prove that

$$\text{res}(f, w) = \lim_{z \to w}(z - w)\left(f(z) - \frac{c}{(z - w)^2}\right).$$

Hence, find the residue at 0 of $\dfrac{1}{z^2 \cos z - 2z^3}$.

(x) Generalize Method B to the case that p has a zero of order k and q has a zero of order $k + 1$.

(xi) Suppose that f has a simple pole at w and g is analytic at w. Prove that $\text{res}(fg, w) = g(w)\,\text{res}(f, w)$. State and prove a generalization of this statement to higher order poles.

(xii) Is the following true or false? 'If f and g have a simple pole at w, then $\text{res}(fg, w) = \text{res}(f, w)\,\text{res}(g, w)$.' Prove or give a counterexample.

(xiii) Find the residue of $\sin(z)\sin(1/z)$ at 0.

(xiv) Show that if f is an even function with a pole at 0, then $\operatorname{res}(f, 0) = 0$.

(xv) Suppose that f has a pole at w. Show that the residue of f' at w is zero.

(xvi) Let D be an open set such that $z \in D \implies \bar{z} \in D$. Suppose that $f : D \to \mathbb{C}$ is a complex function such that $f(\bar{z}) = \overline{f(z)}$.

    (a) Prove that if f has a pole of order N at w, then f has a pole of order N at $\bar{w}$.

    (b) Show that $\operatorname{res}(f, \bar{w}) = \overline{\operatorname{res}(f, w)}$.

This shows that in certain circumstances poles come in conjugate pairs.

(xvii) Suppose that $f(z) = \dfrac{p(z)}{q(z)}$ where p and q are analytic at w with $p(w) \neq 0$, $q(w) = q'(w) = q''(w) = 0$ and $q'''(w) \neq 0$. Show that f has a pole of order 3 at 0 and find a formula for the residue.

Generalise this result to: Suppose that $f(z) = \dfrac{p(z)}{q(z)}$ where p and q are analytic at w with $p(w) \neq 0$, $q(w) = q'(w) = \cdots = q^{(k-1)}(w) = 0$ and $q^{(k)}(w) \neq 0$. Then f has a pole of order k at 0 and the residue $\operatorname{res}(f, w)$ is calculated by

$$\left(\frac{k!}{q^{(k)}(w)}\right)^k \det \begin{vmatrix} \dfrac{q^{(k)}(w)}{k!} & 0 & \cdots & 0 & \dfrac{p(w)}{0!} \\[2ex] \dfrac{q^{(k+1)}(w)}{(k+1)!} & \dfrac{q^{(k)}(w)}{k!} & \cdots & 0 & \dfrac{p'(w)}{1!} \\[2ex] \vdots & \vdots & \ddots & \vdots & \vdots \\[2ex] \dfrac{q^{(2k-1)}(w)}{(2k-1)!} & \dfrac{q^{(2k-2)}(w)}{(2k-2)!} & \cdots & \dfrac{q^{(k+1)}(w)}{(k+1)!} & \dfrac{q^{(k-1)}(w)}{(k-1)!} \end{vmatrix}.$$

Hint: Like in Theorem 17.9 prove a formula for $q^{(k)}$ in terms of $r^{(k-2)}$ this time using the general Leibniz rule for higher derivatives of a product.

(xviii) Suppose that $f(z) = 1/q(z)$ where q is analytic and has a zero of order 3. Conjecture a formula for $\operatorname{res}(f, w)$ and show that it holds.

Summary

❑ The residue of f at z_0 is the coefficient a_{-1} in the Laurent expansion $\sum_{n=-\infty}^{\infty} a_n(z - z_0)^n$. This is denoted $\text{res}(f, z_0)$.

❑ Simple pole: $\text{res}(f, w) = \lim_{z \to w}(z - w)f(z)$.

❑ Some quotients: Suppose $p(w) \neq 0$, $q(z) = 0$ and $q'(w) \neq 0$. Then,

$$\text{res}\left(\frac{p}{q}, w\right) = \frac{p(w)}{q'(w)}.$$

❑ Pole of order N:

$$\text{res}(f, w) = \frac{1}{(N - 1)!} \lim_{z \to w} \frac{d^{N-1}}{dz^{N-1}} \left\{(z - w)^N f(z)\right\}.$$

❑ Calculate expansions directly, equate coefficients, etc.

❑ Suppose that $f(z) = \dfrac{p(z)}{q(z)}$ where p and q are analytic at w, $p(w) \neq 0$, $q(w) = q'(w) = 0$ and $q''(w) \neq 0$. Then f has a double pole at w and

$$\text{res}(f, w) = \frac{2}{(q''(w))^2} \det \begin{vmatrix} p'(w) & p(w) \\ \frac{1}{3}q'''(w) & q''(w) \end{vmatrix}.$$

CHAPTER 18

Cauchy's Residue Formula

In this chapter we calculate certain complex integrals by merely calculating residues and winding numbers. The power of the central theorem, Cauchy's Residue Formula, cannot be overestimated. Calculating integrals is hard but is something we would like to do, for example to solve ordinary differential equations. This theorem says we can do it by calculating residues. From the previous chapter we know we can often calculate these using differentiation, a process much easier than integration.

Theorem 18.1 (Cauchy's Residue Formula)
Let D be an open set in $\mathbb{C}$ and γ be a closed contour such that $\gamma^* \cup Int(\gamma) \subseteq D$. Let $\{p_1, \ldots, p_m\} \subseteq D \backslash \gamma^*$ and $f : D \to \mathbb{C}$ be analytic on D except for poles at $p_1, \ldots, p_m$. Then

$$\int_\gamma f(z)\, dz = 2\pi i \sum_{j=1}^{m} n(\gamma, p_j)\, \text{res}(f, p_j).$$

Proof. We can assume that all the poles are in the interior of γ as then $n(\gamma, p_j) \neq 0$. Let C_j be a circle contour of radius $\varepsilon > 0$ at p_j, taken anticlockwise. Then, there exists $\varepsilon > 0$ such that $C_j \subseteq \gamma^* \cup Int(\gamma)$ and $C_j \cap C_l = \emptyset$ for all j and l with $j \neq l$ and so that C_j lies within the disc upon which the Laurent expansion for f at p_j holds.

There exist contours β_j from some point on the image of γ to the start of C_j. We can define γ_j as the part of γ from the start of β_j to β_{j+1}. So $\gamma = \gamma_1 + \cdots + \gamma_m$.

This set up is shown in Figure 18.1.

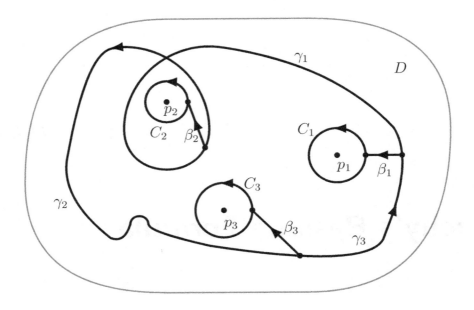

Figure 18.1: Cutting out the poles

Next, we define a contour $\widetilde{\gamma}$ by

$$
\begin{aligned}
\widetilde{\gamma} \;=\; & \beta_1 - n(\gamma, p_1)C_1 - \beta_1 + \gamma_1 \\
& + \beta_2 - n(\gamma, p_2)C_2 - \beta_2 + \gamma_2 + \dots \\
& + \beta_m - n(\gamma, p_m)C_m - \beta_m + \gamma_m.
\end{aligned}
$$

As f is analytic on $\mathrm{Int}(\widetilde{\gamma})$ by Cauchy's Theorem we have $\int_{\widetilde{\gamma}} f = 0$. But, noting that the integrals over β and $-\beta$ cancel,

$$
\int_{\widetilde{\gamma}} f = \sum_{j=1}^{m} -n(\gamma, p_j) \int_{C_j} f + \sum_{j=1}^{m} \int_{\gamma_j} f
$$

$$
0 = \left(\sum_{j=1}^{m} -n(\gamma, p_j) \int_{C_j} f \right) + \int_{\gamma} f.
$$

The result will follow if $\int_{C_j} f = 2\pi i \operatorname{res}(f, p_j)$. This we now show.

The radius of C_j, ε, was chosen so that C_j^* lies within a neighbourhood of p_j

where the function f has a Laurent expansion $f(z) = \sum_{n=-N}^{\infty} a_n(z-p_j)^n$. So,

$$\int_{C_j} f(z)\, dz = \int_{C_j} \left(\sum_{n=-N}^{-1} a_n(z-p_j)^n + \sum_{n=0}^{\infty} a_n(z-p_j)^n \right) dz$$

$$= \int_{C_j} \sum_{n=-N}^{-1} a_n(z-p_j)^n\, dz + \int_{C_j} \sum_{n=0}^{\infty} a_n(z-p_j)^n\, dz$$

$$= \sum_{n=-N}^{-1} \int_{C_j} a_n(z-p_j)^n\, dz + 0.$$

The last equality follows from Theorem 16.18 and Cauchy's Theorem. By the Fundamental Example,

$$\int_{C_j} a_n(z-p_j)^n\, dz = a_n \int_{C_j} (z-p_j)^n\, dz$$

$$= \begin{cases} a_{-1} \times 2\pi i, & n = -1, \\ 0, & n \neq -1. \end{cases}$$

Thus, $\int_{C_j} f(z)\, dz = 2\pi i a_{-1} = 2\pi i \operatorname{res}(f, p)$ as required. $\qquad \square$

There we have it, one of the best theorems in mathematics! It is great because it allows, via residues, straightforward calculation of complicated integrals.

Examples 18.2

(i) Evaluate the integral $\int_{\gamma} \dfrac{dz}{\sin z}$, where γ is a diamond with vertices $5, 7i, -5, -7i$ oriented anticlockwise.

The function $f(z) = \dfrac{1}{\sin z}$ has poles where $\sin z = 0$, i.e., $z = \pi n$, with $n \in \mathbb{Z}$. Of these, $z = 0$ and $\pm \pi$ lie inside γ and have winding number equal to 1. The pole at any point πn is a simple pole with residue

$$\left. \frac{1}{\cos z} \right|_{z=\pi n} = (-1)^n$$

by Method B. So

$$\int_{\gamma} \frac{1}{\sin z}\, dz = 2\pi i \left(\operatorname{res}(f, -2\pi i) + \operatorname{res}(f, 0) + \operatorname{res}(f, 2\pi i) \right)$$

$$= 2\pi i \times (-1 + 1 + (-1))$$

$$= -2\pi i.$$

(ii) Evaluate the integral $\displaystyle\int_\gamma \frac{1}{z^2 + (i-1)z - i}\,dz$, where γ is a square with sides of length 4 centred at the origin, oriented anti-clockwise.

Solution: Let $f(z) = \dfrac{1}{z^2 + (i-1)z - i} = \dfrac{1}{(z+i)(z-1)}$. Therefore, f has simple poles at $z = 1$ and $z = -i$.

Let $q(z) = z^2 + (i-1)z - i$, then $q'(z) = 2z + i - 1$. We have $q'(1) = 2 + i - 1 = 1 + i$, and $q'(-i) = -2i + i - 1 = -1 - i$. Thus,

$$\begin{aligned}
\operatorname{res}(f, 1) &= \frac{1}{1+i} = \frac{1-i}{2} \\
\text{and } \operatorname{res}(f, -i) &= \frac{1}{-1-i} = \frac{-1+i}{2}.
\end{aligned}$$

We have $n(\gamma, 1) = n(\gamma, -i) = 1$, so by Cauchy's Residue Theorem

$$\begin{aligned}
\int_\gamma \frac{1}{z^2 + (i-1)z - i}\,dz &= 2\pi i\left(\operatorname{res}(f, 1) + \operatorname{res}(f, -i)\right) \\
&= 2\pi i\left(\frac{1-i}{2} + \frac{-1+i}{2}\right) \\
&= 0.
\end{aligned}$$

The next theorem gives a surprising relationship between the roots of a polynomial.

Theorem 18.3
Suppose that p is a complex polynomial of degree $n \geq 2$ with distinct roots, α_1, α_2, ..., α_n. Then,

$$\sum_{j=1}^n \frac{1}{p'(\alpha_j)} = 0.$$

Proof. The function $f(z) = \dfrac{1}{p(z)}$ has a simple pole at α_j for all j and hence $\operatorname{res}(f, \alpha_j) = 1/p'(\alpha_j)$ by Method B. Let C_R be a circle of radius R centred at 0 as usual. From Cauchy's Residue Theorem we have that

$$\int_{C_R} f(z)\,dz = 2\pi i \sum_{j=1}^n \operatorname{res}(f, \alpha_j) = 2\pi i \sum_{j=1}^n \frac{1}{p'(\alpha_j)}.$$

Now let's use the Estimation Lemma to show that the integral is zero.

By a consequence of the Fundamental Theorem of Algebra, the polynomial p can be written as

$$p(z) = c \prod_{j=1}^{n} (z - \alpha_j),$$

where $c \in \mathbb{C}$. Thus for R with all the roots of p contained in the interior of C_R we have

$$|p(z)| = |c| \prod_{j=1}^{n} |z - \alpha_j| \geq |c| \prod_{j=1}^{n} (R - |\alpha_j|).$$

Hence for large enough R we have by the Estimation Lemma that

$$\left| \int_{C_R} f(z) \, dz \right| \leq \frac{2\pi R}{|c| \prod_{j=1}^{n} (R - |\alpha_j|)}.$$

Taking the limit as $R \to \infty$ and as $n \geq 2$ we deduce that $\int_{C_R} f(z) \, dz = 0$. $\qquad \square$

From this we can produce an interesting result about real functions.

Corollary 18.4
Suppose that p is a real polynomial of degree $n \geq 2$, such that all roots, $\alpha_1, \alpha_2, \ldots, \alpha_n$, are real and distinct. Then,

$$\sum_{j=1}^{n} \frac{1}{p'(\alpha_j)} = 0.$$

Proof. Follows simply from the theorem. $\qquad \square$

By requiring that p and all its roots are real we can give geometrical meaning to $1/p'(\alpha_j)$. It is the reciprocal of the slope of the graph of p as it cuts the x-axis. Who would have guessed that the sum of these is zero?

Argument Theorem

For some integrals we don't even need to calculate the residues we need only find the multiplicity of zeros and poles.

Theorem 18.5 (Argument Theorem)
Let γ be a closed simple contour such that $\gamma^* \cup \text{Int}(\gamma) \subseteq D$ for some open set D. Let f have a finite number of zeros and let f be analytic on D except for a finite number poles inside γ. Then

$$\frac{1}{2\pi i} \int_{\gamma} \frac{f'(z)}{f(z)} dz = Z_f(\gamma) - P_f(\gamma)$$

where $Z_f(\gamma)$ is the number of zeros and $P_f(\gamma)$ is the number of poles inside γ, each counted with multiplicity.

Proof. Let z_0 be a point such that $f(z) = (z - z_0)^n g(z)$ for some non-zero integer n and analytic function g with $g(z_0) \neq 0$. Then

$$f'(z) = n(z - z_0)^{n-1} g(z) + (z - z_0)^n g'(z)$$

and so

$$\frac{f'(z)}{f(z)} = \frac{n}{z - z_0} + \frac{g'(z)}{g(z)}.$$

Since $g(z_0) \neq 0$, then

$$\operatorname{res}\left(\frac{f'}{f}, z_0\right) = n.$$

Now if z_0 is a zero of f inside γ then n is a positive integer, if z_0 is a pole of f, then n is a negative integer. Hence, by the residue theorem

$$\frac{1}{2\pi i} \int \frac{f'(z)}{f(z)} dz = \frac{1}{2\pi i} 2\pi i \left(\sum_{\text{zeros } z_0} \operatorname{res}\left(\frac{f'}{f}, z_0\right) + \sum_{\text{poles } z_0} \operatorname{res}\left(\frac{f'}{f}, z_0\right) \right)$$

$$= \sum_{\text{zeros } z_0} \operatorname{order}(f, z_0) - \sum_{\text{poles } z_0} \operatorname{order}(f, z_0)$$

$$= Z - P.$$

$\square$

Example 18.6
Let γ describe the circle of radius 10. Calculate

$$\int_\gamma \frac{z^2 + 1}{3z^3 + 3z + 1} dz.$$

To solve this consider $f(z) = z^3 + 3z + 1$. Then, $Z_f = 3$ and $P_f = 0$ since f has all its three zeros within γ and no poles.

$$\int_\gamma \frac{z^2 + 1}{3z^3 + 3z + 1} dz = \frac{1}{3} \int_\gamma \frac{3z^2 + 3}{3z^3 + 3z + 1} dz$$

$$= \frac{1}{3} \int_\gamma \frac{(3z^3 + 3z + 1)'}{3z^3 + 3z + 1} dz$$

$$= \frac{1}{3} \times 2\pi i (Z_f - P_f)$$

$$= \frac{1}{3} \times 2\pi i (3 - 0)$$

$$= 2\pi i.$$

Now we need to explain why the theorem is called the Argument Theorem. Let $\zeta = f(z)$, then

$$\frac{1}{2\pi i} \int_\gamma \frac{f'(z)}{f(z)} dz = \frac{1}{2\pi i} \int_{f\circ\gamma} \frac{1}{\zeta} d\zeta.$$

Thus, the integral is the winding number of the contour $f\circ\gamma$ around the origin and a winding number is just the total increase in the argument of $f(z)$ for $z = \gamma(t)$.

Example 18.7
Let $f(z) = \dfrac{z}{(z^2+2)^2}$. Let γ_2 denote the circle of radius 2 about i and γ_4 denote the circle of radius 4 about i. Calculate the winding number about 0 of $f \circ \gamma_2$ and $f \circ \gamma_4$.

First, note that f has a zero of multiplicity 1 at the origin and poles at $z = \pm\sqrt{2}i$ of multiplicity 2. Note that γ_2 contains the origin and the pole at $\sqrt{2}i$ whereas γ_4 contains all the zeros and poles of f.

Therefore,

$$n(f \circ \gamma_2, 0) = \frac{1}{2\pi i} \int_{\gamma_2} \frac{f'(z)}{f(z)} dz = 1 - 1 \times 2 = -1,$$

and

$$n(f \circ \gamma_4, 0) = \frac{1}{2\pi i} \int_{\gamma_4} \frac{f'(z)}{f(z)} dz = 1 - 2 \times 2 = -3.$$

Rouché's Theorem

The following theorem allows us to say something about the number of zeros within a curve.

Theorem 18.8 (Rouché's Theorem)
Let γ be a simple closed contour. Suppose that f and g are analytic on set containing $\gamma^* \cup \mathrm{Int}(\gamma)$ and that $|g(z)| < |f(z)|$ for all $z \in \gamma^*$. Then, f and $f + g$ have the same number of zeros in $\mathrm{Int}(\gamma)$. (Zeros counted with multiplicity.)

Proof. Let $\Delta_\gamma \arg(h)$ denote the increase in $\arg(h)$ for a function h as we traverse γ once in the positive direction. (For the sake of clarity we need to introduce this notation. See Exercise (xvi) at the end of the chapter.) From the Argument Theorem we know that

$$\Delta_\gamma \arg(h) = 2\pi Z_h$$

where Z_h is the number of zeros of h within γ. Thus, $\Delta_\gamma \arg(f) = 2\pi Z_f$ and $\Delta_\gamma \arg(f + g) = 2\pi Z_{f+g}$.

We have

$$2\pi Z_{f+g} = \Delta_\gamma \arg(f + g)$$

$$= \Delta_\gamma \arg\left(f\left(1 + \frac{g}{f}\right)\right), \quad \text{(why is } f(z) \neq 0?)$$

$$= \Delta_\gamma \arg(f) + \Delta_\gamma \arg\left(1 + \frac{g}{f}\right)$$

$$= 2\pi Z_f + \Delta_\gamma \arg\left(1 + \frac{g}{f}\right).$$

Therefore, we need to show that $\Delta_\gamma \arg\left(1 + \dfrac{g}{f}\right) = 0$.

If the contour defined by $1 + g(\gamma(t))/f(\gamma(t))$ winds round 0, then for some t its real part is negative. But,

$$\text{Re}\left(1 + \frac{g(\gamma(t))}{f(\gamma(t))}\right) > 1 - \left|\frac{g(\gamma(t))}{f(\gamma(t))}\right| > 0$$

as $|g(z)| < |f(z)|$ on γ^*. $\qquad\qquad\qquad\square$

Example 18.9

Consider the polynomial $p(z) = 2z^4 - 7z - 3$. Let $f(z) = -7z$ and $g(z) = 2z^4 - 3$. Then, for $|z| = 1$ we have

$$|g(z)| \leq |2z^4| + |-3| = 2|z|^4 + 3 = 5 < 7 = |-7z^4| = |f(z)|.$$

The function f has one zero within the unit circle so p does too.

Now, let's take $f(z) = 2z^4$ and $g(z) = -7z - 3$. Then, for $|z| = 2$ we have

$$|g(z)| \leq 7|z| + 3 = 17 < 32 = 2 \times 2^4 = 2|z|^4 = |f(z)|.$$

Thus, as f has four zeros within the circle of radius of radius 2, we know that p must too.

The zeros are not on $|z| = 1$. If $2z^4 - 7z - 3 = 0$, then $2z^3 = 7z + 3$. Therefore, for $|z| = 1$, we have

$$2 = 2|z|^4 = |7z + 3| \geq |7|z| - 3| = |7 - 3| = 4.$$

From these results we know that there are three roots of p within the annulus $1 < |z| < 2$.

Example 18.10

Show that $z^4 + 5 = e^z$ has 2 solutions in the left half-plane.

This is equivalent to showing that $z^4 + 5 - e^z$ has 2 roots in the left half-plane. Let $f(z) = z^4 + 5$ and $g(z) = -e^z$. Consider a large semi-circle of radius R in the left half-plane (as in Example 7.7).

On $|z| = R$ for large enough R we have

$$|f(z)| = |z^4 + 5| \geq |z|^4 - 5 = R^4 - 5,$$
$$\text{and } |g(z)| = |-e^z| = e^{\text{Re}(z)} \leq 1.$$

Hence for large R, we have $|g(z)| < |f(z)|$ for $|z| = R$.

On $z = iy$ where $y \in [-R, R]$, we have

$$|f(z)| = |(iy)^4 + 5| \geq |y^4 + 5| \geq 5,$$
$$\text{and } |g(z)| = |-e^{iy}| = 1.$$

Hence, again $|g(z)| < |f(z)|$.

Therefore, within this semi-circle f and $f + g$ have the same number of zeros. The function f has 2 zeros in the left half-plane so, for a large enough semi-circle, f has 2 zeros and the conditions of Rouché's Theorem hold, hence $z^4 + 5 = e^z$ has 2 solution in the left half-plane.

Exercises

Exercises 18.11

(i) Calculate the following integrals. Unless otherwise stated all contours traced anticlockwise. (You may find answers to Exercises 17.14(i) helpful.)

(a) $\displaystyle\int_{\gamma_1} \frac{e^{iz}}{z^3 - 1} \, dz$, where γ_1 describes the boundary of the semicircle $\{|z| \leq 10, \text{Re}(z) \geq 0\}$.

(b) $\displaystyle\int_{\gamma_2} \frac{\cos \pi e^{\pi z}}{z(z + 2i)^2} \, dz$, where γ_2 is the unit circle.

(c) $\displaystyle\int_{\gamma_3} \frac{\cos \pi e^{\pi z}}{z(z + 2i)^2} \, dz$, where γ_3 is the diamond with vertices $1, 10i, -1, -10i$.

(d) $\displaystyle\int_{\gamma_4} \frac{3z^4 + 2z^2 + \cos^2(2z^2)}{z^5} \, dz$, where $\gamma_4(t) = 2 + 3e^{it}$ for $0 \leq t \leq 4\pi$.

(e) $\displaystyle\int_{\gamma_5} \frac{\cos 2z}{z^3} \, dz$, where γ_6 is the square with vertices $\pm 1 \pm i$.

(f) $\int_{\gamma_6} \dfrac{e^{iz}}{z^2 + 4}\, dz$, where γ_6 describes the boundary of the semicircle $\{|z| \leq 10, \operatorname{Im}(z) \geq 0\}$.

(g) $\int_{\gamma_2} \dfrac{e^{\sin z}}{z(z-2)^2}\, dz$, where γ_2 is the unit circle.

(h) $\int_{\gamma_3} \dfrac{dz}{e^z - 1}$, where γ_3 is as above.

(ii) For $f : \mathbb{C}\backslash P \to \mathbb{C}$ be a continuous map where P is a finite number points. Let $\varphi : (0, \infty)\backslash\{|z| : z \in P\} \to \mathbb{R}$ be defined by

$$\varphi(r) = \left| \int_{C_r} f(z)\, dz \right|$$

where C_r is our standard contour giving the circle of radius r centred at the origin.

Sketch the graph of φ for each of the following functions. (These also appear in Exercises 17.14(i).)

(a) $\dfrac{e^z}{z^6(z-1)}$,   (b) $\dfrac{\sin z}{z^6(z-1)}$,   (c) $\dfrac{\cos(\pi e^{\pi z})}{z(z+2i)^2}$,   (d) $\cot z$.

(iii) Find $\int_{\gamma} \left(z + \dfrac{1}{z} \right)^n dz$ where γ is a contour.

(iv) Deduce Cauchy's Integral Formula from Cauchy's Residue Formula.

(v) Generalize Theorem 18.3 to the case of non-distinct roots of a polynomial.

(vi) What is the sum of the residues of the function $\dfrac{1}{z^{120} + 3z^{17} + 5}$? (Hint: use Cauchy's Residue Formula and the Estimation Lemma.)

Generalize this result.

(vii) The Residue Theorem can be used to find a **partial fraction representation** of a function.

Prove that if f is a complex differentiable function on $\mathbb{C}$ except at a finite number of simple poles, $p_1, p_2, \ldots, p_m$, and there exists positive real numbers M and R such that $|f(z)| \leq M$ for all $|z| > R$, then there exists $c \in \mathbb{C}$ such that

$$f(z) = c + \sum_{j=1}^{m} \frac{\operatorname{res}(f, p_j)}{z - p_j}$$

for all $z \neq p_j$, where $c = \lim\limits_{|z| \to \infty} f(z)$.

(viii) Using the previous exercise find partial fraction representations for

(a) $f(z) = \dfrac{(z+2)(z+1)}{(z-1)(z-3)(z-5)}$,

(b) $f(z) = \dfrac{z^2+3}{(z-1)(z-2)}$.

(ix) Let's generalize the formula for a_n in Taylor's Theorem: Show that

$$a_n = \frac{1}{2\pi i} \int_\gamma \frac{f(z)}{(z-z_0)^{n+1}}\, dz,$$

for all closed contours $\gamma : [a, b] \to D$ of winding number 1 about z_0.

(x) By considering the zeros of z^n reprove the Fundamental Theorem of Algebra with the stronger conclusion that the number of zeros is equal to the degree of the polynomial.

(xi) Suppose that f is analytic on a neighbourhood of z_0 and has a zero of multiplicity m there. Let g be any analytic function. Show that there exists $t > 0$ and $r > 0$ such that $f + tg$ has m zeros inside C_r, a circle of radius r centred at z_0.

This means that under small perturbations the zero of a function may break up into other zeros but the total multiplicity remains the same. (For example, let $f(z) = z^2(z-3)$, then for small t, $f + tz$ has two zeros of multiplicity 1 near the origin.) This is called **conservation of multiplicity**.

(xii) Find the number of roots of the following equations inside the circle of radius 1.

(a) $z^5 + 8z + 1 = 0$,

(b) $z^7 + 8z^3 + 3z = 0$.

How many zeros lie within $|z| = 2$?

(xiii) How man zeros does $5z^4 - 5iz^3 + 37z^2 - iz + 14$ have in upper half plane?

(xiv) Show that $z^5 + 2iz^3 - 3z^2 + 18iz + 11 = 0$ has 4 solutions such that $|z| \geq 1$.

(xv) Generalize the Argument Theorem as follows: Let $f : D \to \mathbb{C}$ be an analytic function on the open set D except at a finite number of poles. Let γ be a closed simple contour such that $\gamma^* \cup \text{Int}(\gamma) \subseteq D$. Let $\{z_1, \ldots, z_m\}$ be the set of zeros of f that lie in the interior of γ and let $\{p_1, \ldots, p_n\}$ be the set of poles of f that lie in the interior of γ. Show that if $g : D \to \mathbb{C}$ is an analytic function, then

$$\frac{1}{2\pi i} \int_\gamma \frac{f'}{f} \, dz = \sum_{j=1}^m \text{order}(f, z_j) g(z_j) + \sum_{j=1}^n \text{order}(f, p_j) g(p_j)$$

where $\text{order}(w)$ is the multiplicity of the zero or pole at w.

(xvi) In the proof of Rouché's Theorem we used $\Delta_\gamma \arg(h)$ rather than $n(h \circ \gamma, 0)$ for denoting the increase in the argument of $h \circ \gamma$. This is to avoid the following mistake. Recalling that the number of zeros of h within γ is given by $n(h \circ \gamma, 0)$ consider following: Suppose that f and g are analytic on set containing $\gamma^* \cup \text{Int}(\gamma)$ for a closed simple contour γ. Then

$$Z_{f+g} = n\left((f + g) \circ \gamma, 0\right) = n\left(f \circ \gamma + g \circ \gamma, 0\right)$$
$$= n\left(f \circ \gamma, 0\right) + n\left(g \circ \gamma, 0\right) = Z_f + Z_g.$$

Why is this argument false? (Hint: See Remark 6.5.)

Summary

❏ Let D be domain and γ a closed contour such that $\gamma^* \cup \text{Int}(\gamma) \subseteq D$. Let $\{p_1, \ldots, p_m\} \subseteq D \backslash \gamma^*$ and $f : D \to \mathbb{C}$ be analytic with poles at $p_1, \ldots, p_m$. Then

$$\int_\gamma f(z) \, dz = 2\pi i \sum_{j=1}^m n(\gamma, p_j) \, \text{res}(f, p_j).$$

❏ Argument Theorem: Let γ be a closed simple contour such that $\gamma^* \cup \text{Int}(\gamma) \subseteq D$ for some open set D. Let f have a finite number of zeros and let f be analytic on D except for a finite number poles inside γ. Then

$$\frac{1}{2\pi i} \int_\gamma \frac{f'(z)}{f(z)} dz = Z_f(\gamma) - P_f(\gamma)$$

where $Z_f(\gamma)$ is the number of zeros and $P_f(\gamma)$ is the number of poles inside γ, each counted with multiplicity.

CHAPTER 19

Evaluation of Real Integrals

The next few chapters demonstrate the power of complex analysis by applying it to different areas of mathematics that initially appear to have no connection to it:

❏ definite real integrals,

❏ integrals of periodic functions,

❏ summation of series.

In each case complex analysis allows us to solve, in a simple manner, problems involving real functions that are not easy to solve with only real methods. It is here that our hard work really pays off and the reason for extending calculus to the complex domain becomes apparent. Furthermore, there are many other topics that benefit from the use of complex analysis, for example, Fourier series, number theory and solutions of differential equations via Laplace Transforms.

The basic idea

We begin with evaluating integrals of the form

$$\int_{-\infty}^{\infty} f(x)\, dx \qquad \text{and} \qquad \int_{0}^{\infty} f(x)\, dx$$

where f is a real function.

The basic method involves a number of steps, none of which is particularly complicated, but the method at first appears rather involved. Nonetheless, it is worth persevering due to the impressive results we can achieve with it and its variations.

First we will define a contour via the sum of two contours. Let R be a positive real number. Define the following:

$$\gamma_R(t) = t, \text{ where } -R \le t \le R,$$
$$C_R^+(t) = Re^{it}, \text{ where } 0 \le t \le \pi,$$
$$\gamma = \gamma_R + C_R^+.$$

Note that C_R^+ is just the positive imaginary part our usual C_R. These contours can be seen in the following diagram.

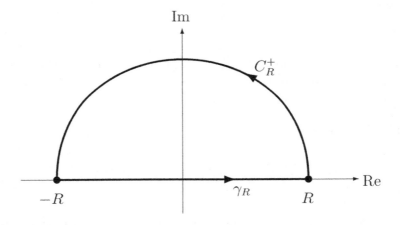

Now assume that we can extend the real function given by $f(x)$ to a function on $\mathbb{C}$ which perhaps has a few isolated singularities. As usual we'll use z to denote the complex variable. Integrating f round the contour $\gamma = \gamma_R + C_R^+$ we see that

$$\int_\gamma f(z)\,dz = \int_{\gamma_R} f(z)\,dz + \int_{C_R^+} f(z)\,dz. \tag{19.1}$$

The left-hand side is a contour integral of a closed contour and so we can calculate it using Cauchy's Residue Formula.

On the right-hand side, consider the integral of f over γ_R. We have $\gamma_R'(t) = 1$ for $-R \le t \le R$ and so,

$$\int_{\gamma_R} f(z)dz = \int_{-R}^{R} f(t).1\,dt = \int_{-R}^{R} f(t)\,dt.$$

If we let $R \to \infty$, then often we obtain the real integral $\int_{-\infty}^{\infty} f(x)\,dx$, i.e., the integral we wish to calculate. (We'll say what we mean by 'often' later.)

We will show that in many cases the other term on the right-hand side of equation 19.1 tends to zero. Hence, the limit of the real integral we wish to calculate is equal to the integral calculated via residues.

Let's summarize the proceeding as the following, under some (as yet unstated) restrictions, we have

$$\underbrace{\int_{\gamma} f(z)\,dz}_{\text{Calculate via residues}} = \underbrace{\int_{\gamma_R} f(z)\,dz}_{\text{tends to } \int_{-\infty}^{\infty} f(x)\,dx} + \underbrace{\int_{C_R^+} f(z)\,dz}_{\text{tends to } 0}.$$

Thus we have a method for determining a real integral from residues! Let's do an example to make this clear.

Example 19.1
Find the integral $\displaystyle\int_{-\infty}^{\infty} \frac{dx}{(x^2+1)^2}$.

(i) Let R be a positive real number. Let $\gamma_R(t) = t$ where $-R \le t \le R$ and $C_R^+(t) = Re^{it}$ where $0 \le t \le \pi$. Let $\gamma = \gamma_R + C_R^+$.

(ii) Let $f(z) = \dfrac{1}{(z^2+1)^2}$. Then $f(z) = \dfrac{1}{(z^2+1)^2} = \dfrac{1}{(z-i)^2(z+i)^2}$. So f has poles of order 2 at $z = i$ and $z = -i$. Only the $z = i$ pole lies in the interior of γ (when $R > 1$).

By Method C we have

$$
\begin{aligned}
\mathrm{res}(f, i) &= \frac{1}{(2-1)!} \lim_{z \to i} \frac{d}{dz}\left((z-i)^2 \frac{1}{(z-i)^2(z+i)^2} \right) \\
&= \lim_{z \to i} \frac{d}{dz}\left(\frac{1}{(z+i)^2} \right) \\
&= \left. \left(-2\frac{1}{(z+i)^3} \right) \right|_{z=i} \quad \text{as the function is continuous at } z = i, \\
&= -2\frac{1}{(i+i)^3} \\
&= \frac{1}{4i}.
\end{aligned}
$$

(iii) To show that the integral of $f(z)$ along C_R^+ tends to 0 as $R \to \infty$ we can use the Estimation Lemma. For $z \in \gamma_R^*$, we have

$$|z^2 + 1| \geq \left| |z|^2 - 1 \right|$$
$$= R^2 - 1, \text{ for } R > 1,$$
$$|z^2 + 1|^2 \geq (R^2 - 1)^2$$
$$\frac{1}{(R^2 - 1)^2} \geq \frac{1}{|(z^2 + 1)^2|}.$$

Thus,

$$\left| \int_{C_R^+} \frac{1}{(z^2 + 1)^2}\, dz \right| \leq \frac{\pi R}{(R^2 - 1)^2} \to 0 \text{ as } R \to \infty.$$

Hence,

$$\int_{C_R^+} \frac{1}{(z^2 + 1)^2}\, dz \to 0 \text{ as } R \to \infty.$$

(iv) Therefore, for $R > 1$,

$$\int_\gamma f(z)\, dz = \int_{\gamma_R} f(z)\, dz + \int_{C_R^+} f(z)\, dz$$

$$2\pi i \operatorname{res}(f, i) = \int_{-R}^{R} \frac{1}{(x^2 + 1)^2}\, dx + \int_{C_R^+} f(z)\, dz$$

$$2\pi i \left(\frac{1}{4i} \right) = \int_{-\infty}^{\infty} \frac{1}{(x^2 + 1)^2}\, dx + 0 \text{ as } R \to \infty \qquad \text{(see remark below)}$$

$$\frac{\pi}{2} = \int_{-\infty}^{\infty} \frac{1}{(x^2 + 1)^2}\, dx$$

Remark 19.2
The method we have given does not always calculate the integral we want! The integral $\int_{-\infty}^{\infty} f(x)\, dx$ is *defined* to be

$$\lim_{a \to -\infty} \lim_{b \to \infty} \int_a^b f(x)\, dx$$

and in fact, may *not* equal

$$\lim_{R \to \infty} \int_{-R}^{R} f(x)\, dx$$

which is in fact what we have calculated. The latter is called the **principal value of the integral** and is denoted PV $\int_{-\infty}^{\infty} f(x)\, dx$.

The principal value of an integral may exist even if $\int_{-\infty}^{\infty} f(x)\, dx$ does not. For example, it is easy to show that $\int_{-\infty}^{\infty} x\, dx$ does not exist but PV $\int_{-\infty}^{\infty} x\, dx = 0$. However, if both integrals exist, then they are equal.

Proposition 19.3

Let f be a continuous real function. If $|f(x)| \leq C/x^2$ for large $|x|$ and some C, then $\int_{-\infty}^{\infty} f(x)\,dx$ exists.

This can be proved using the methods of real analysis.

Before we use this proposition in an example let's find a condition that implies the conditions of Proposition 19.3 and, in addition, allows us to show that the integral on the semi-circle tends to zero as the semi-circle gets larger.

Jordan's Lemma (Weak Version)

We can use the Estimation Lemma to show that the integral on the semi-circle tends to zero, however the following lemma avoids the tedious calculations.

Lemma 19.4 (Jordan's Lemma (Weak Version))

Suppose that p and q are polynomials with $\deg(p) \leq \deg(q) - 2$. Let $C_R^+(t) = Re^{it}$, for $0 \leq t \leq \pi$ be the semi-circular contour of radius R and let $a \geq 0$.

Then,

$$\int_{C_R^+} \frac{p(z)}{q(z)} e^{iaz}\,dz \to 0 \text{ as } R \to \infty.$$

Proof. As usual, we apply the Estimation Lemma. We have,

$$\left| e^{iaz} \right| = e^{\text{Re}(iaz)} = e^{\text{Re}(ia(x+iy))} = e^{\text{Re}(-ay+iax)} = e^{-ay} \leq 1,$$

the latter inequality holds because $y \geq 0$ for $z \in C_R^{+*}$, and because $a \geq 0$.

Now,

$$p(z) = \sum_{k=0}^{\deg p} a_k z^k$$

and so, for $z \in C_R^{+*}$ (thus $|z| = R$), we have

$$|p(z)| \leq \sum_{k=0}^{\deg p} \left| a_k z^k \right| = \sum_{k=0}^{\deg p} |a_k|\,|z|^k = \sum_{k=0}^{\deg p} |a_k|\,R^k.$$

By a corollary of the Fundamental Theorem of Algebra, the polynomial q can be written as

$$q(z) = c \prod_{j=1}^{\deg q} (z - \alpha_j),$$

where c is some constant and α_j is a root of q. We then have for $z \in C_R^{+*}$ and $R \geq \max\{|\alpha_j|\}$ that

$$|q(z)| = |c| \prod_{j=1}^{\deg q} (|z - \alpha_j|) \geq |c| \prod_{j=1}^{\deg q} (R - |\alpha_j|).$$

Note that the right-hand side is a polynomial in R of degree $\deg q$.

Thus, for large enough R, by the Estimation Lemma,

$$\left| \int_{C_R^+} \frac{p(z)}{q(z)} e^{iaz} \, dz \right| \leq \frac{\sum_{k=0}^{\deg p} |a_k| \, R^k}{|c| \prod_{j=1}^{\deg q} (R - |\alpha_j|)} \times \pi R.$$

So, if $\deg p \leq \deg q - 2$, then the right-hand side tends to 0 as $R \to \infty$. Hence, the integral in the left-hand side tends to 0 as $R \to \infty$. $\square$

Remark 19.5

For $f(x) = p(x)e^{iax}/q(x)$, p and q polynomials with $a \geq 0$, it can be show that the condition $\deg(p) \leq \deg(q) - 2$ implies that $|f(x)| \leq C/x^2$ for x real and large for some C. Hence, in this situation $\int_{-\infty}^{\infty} f(x) \, dx$ exists and is equal to PV $\int_{-\infty}^{\infty} f(x) \, dx$.

Example 19.6

Let us show that

$$\int_{-\infty}^{\infty} \frac{x^2}{4x^4 + 10x^2 + 4} \, dx = \frac{\pi}{6\sqrt{2}}.$$

First we find the poles and residues in the upper half plane for the function $f(z) = z^2/(4z^4 + 10z^2 + 4)$. The quartic $4z^4 + 10z^2 + 4$ is quadratic in z^2 and so we can find its roots by solving $4w^2 + 10w + 4 = 0$ where $w = z^2$. This gives

$$w = -\frac{1}{2} \text{ and } w = -2$$

and so

$$z = \pm \frac{i}{\sqrt{2}} \text{ and } z = \pm\sqrt{2}i.$$

The poles in the upper half plane are at $\sqrt{2}i$ and $i/\sqrt{2}$. Both are simple as the root is not repeated. Using Method B we have that

$$\text{res}(f, p) = \frac{p^2}{4p(4p^2 + 5)},$$

so

$$\text{res}(f, \sqrt{2}i) = \frac{(\sqrt{2}i)^2}{4\sqrt{2}i\left(4(\sqrt{2}i)^2 + 5\right)} = \frac{1}{6\sqrt{2}i}$$

and

$$\text{res}(f, i/\sqrt{2}) = \frac{(i/\sqrt{2})^2}{4(i/\sqrt{2})\left(4(i/\sqrt{2})^2 + 5\right)} = -\frac{1}{12\sqrt{2}i}.$$

Let γ, γ_R and C_R^+ be the contours defined earlier. Then for $R > \sqrt{2}$ we have by Cauchy's Residue Formula,

$$\int_\gamma f(z)\,dz = 2\pi i \left(\text{res}(f, \sqrt{2}i) + \text{res}(f, i/\sqrt{2}) \right)$$

$$= 2\pi i \left(\frac{1}{6\sqrt{2}i} - \frac{1}{12\sqrt{2}i} \right)$$

$$= \frac{\pi}{6\sqrt{2}}.$$

We have $\int_{C_R^+} f(z)\,dz \to 0$ by Jordan's Lemma as $R \to \infty$ (and the real integral we wish to calculate equals it principal value).

Thus

$$\lim_{R\to\infty} \int_\gamma f = \lim_{R\to\infty} \int_{\gamma_R} f + \lim_{R\to\infty} \int_{C_R^+} f$$

$$\lim_{R\to\infty} \frac{\pi}{6\sqrt{2}} = \lim_{R\to\infty} \int_{-R}^{R} \frac{x^2}{4x^4 + 10x^2 + 4}\,dx + 0$$

$$\frac{\pi}{6\sqrt{2}} = \int_{-\infty}^{\infty} \frac{x^2}{4x^4 + 10x^2 + 4}\,dx.$$

Let's now calculate an integral of the form $\int_0^\infty f(x)\,dx$.

Example 19.7

Compute the real integral $\displaystyle\int_0^\infty \frac{1}{1 + x^{16}}\,dx$.

Let $f(z) = \dfrac{1}{1 + z^{16}}$. First note that $f(-z) = f(z)$ (that is, f is an even function) and so

$$\int_{-\infty}^{\infty} \frac{1}{1 + x^{16}}\,dx = 2\int_0^\infty \frac{1}{1 + x^{16}}\,dx.$$

Therefore, we can use the semi-circle method as before.

The poles occur at z where $1 + z^{16} = 0$. We have

$$1 + z^{16} = 0 \iff z^{16} = -1 \iff z^{16} = e^{\pi i} \iff z = e^{(2k+1)\pi i/16}$$

for $k = 0, \ldots, 15$. Since we have 16 distinct values for a degree 16 polynomial the resulting poles are all simple. The residues for these are easy to calculate using Method B:

$$
\begin{aligned}
\operatorname{res}(f, e^{(2k+1)\pi i/16}) &= \left. \frac{1}{16z^{15}} \right|_{z=\exp((2k+1)\pi i/16)} \\
&= \left. \frac{z}{16z^{16}} \right|_{z=\exp((2k+1)\pi i/16)} \\
&= \frac{e^{(2k+1)\pi i/16}}{16(-1)} \\
&= -\frac{e^{(2k+1)\pi i/16}}{16}.
\end{aligned}
$$

To apply the semi-circle method we need the sum of the residues of the poles in the upper plane, i.e., for $k = 0, \ldots, 7$. Let $w = e^{i\pi/16}$ (so $w^{16} = -1$). The sum is

$$
\begin{aligned}
\sum_{k=0}^{7} -\frac{e^{(2k+1)\pi i/16}}{16} &= -\frac{w}{16} \sum_{k=0}^{7} w^{2k} \\
&= -\frac{w}{16} \frac{(w^2)^8 - 1}{w^2 - 1} \\
&= \frac{w}{16} \frac{2}{w^2 - 1} \\
&= \frac{1}{16i} \frac{2i}{w - w^{-1}} \\
&= \frac{1}{16i} \frac{2i}{e^{i\pi/16} - e^{-i\pi/16}} \\
&= \frac{1}{16i \sin(\pi/16)}.
\end{aligned}
$$

From Jordan's Lemma we deduce that

$$
\int_{C_R^+} \frac{1}{1 + z^{16}} \, dz \to 0 \text{ as } R \to \infty
$$

and the real integral we seek is equal to its principal value.

Therefore, for $R > 1$,

$$\int_\gamma f(z)\,dz = \int_{\gamma_R} f(z)\,dz + \int_{C_R^+} f(z)\,dz$$

$$2\pi i \sum_{k=0}^{7} \operatorname{res}(f, e^{(2k+1)\pi i/16}) = \int_{-R}^{R} \frac{1}{1+x^{16}}\,dx + \int_{C_R^+} f(z)\,dz$$

$$2\pi i \frac{1}{16i\sin(\pi/16)} = \int_{-\infty}^{\infty} \frac{1}{1+x^{16}}\,dx + 0 \text{ as } R \to \infty$$

$$\frac{\pi}{8\sin(\pi/16)} = \int_{-\infty}^{\infty} \frac{1}{1+x^{16}}\,dx$$

$$\frac{\pi}{16\sin(\pi/16)} = \int_{0}^{\infty} \frac{1}{1+x^{16}}\,dx.$$

Exercise 19.8
Find the integral

$$\int_{0}^{\infty} \frac{1}{1+x^{2n}}\,dx$$

for all $n \in \mathbb{N}$.

Product integrands with sines and cosines

With Jordan's Lemma we can calculate integrals of the form

$$\int_{-\infty}^{\infty} f(x)\cos x\,dx \qquad \text{and} \qquad \int_{-\infty}^{\infty} f(x)\sin x\,dx.$$

Let us try an example that really shows the power of complex analysis.

Example 19.9
Calculate the integral $\displaystyle\int_{-\infty}^{\infty} \frac{\cos x}{x^2 - 4x + 13}\,dx$.

The key here is to choose $f(z) = \dfrac{e^{iz}}{z^2 - 4z + 13}$ rather than $\dfrac{\cos z}{z^2 - 4z + 13}$. Remember that $e^{iz} = \cos z + i\sin z$ so we will get the $\cos x$ in the numerator as required, but we will also get a $\sin x$ term. Surprisingly, this turns out to be a bonus, not a problem – just watch!

(i) Let γ be as in the previous example, so $\gamma = \gamma_R + C_R^+$.
(ii) As usual we want to evaluate

$$\int_\gamma f(z)\,dz$$

via Cauchy's Residue Formula.

We know that e^{iz} has no zeros and no poles, so all the poles arise from zeros of the denominator: $z^2 - 4z + 13$, and

$$z^2 - 4z + 13 = 0 \iff z = \frac{4 \pm \sqrt{16 - 52}}{2} = 2 \pm 3i.$$

Let $R > |2 + 3i| = \sqrt{13}$, so $2 + 3i$ is inside the contour, $2 - 3i$ is outside it. The diagram shows the location of poles and the contour C_R^+.

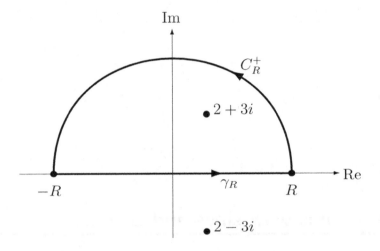

The pole is simple (because it arises from a non-repeated root) so

$$
\begin{aligned}
\int_\gamma \frac{e^{iz}}{z^2 - 4z + 13} \, dz &= 2\pi i \operatorname{res}\left(\frac{e^{iz}}{z^2 - 4z + 13}, 2 + 3i \right) \\
&= 2\pi i \frac{e^{i(2+3i)}}{2(2 + 3i) - 4} \\
&= 2\pi i \frac{e^{-3+2i}}{6i} \\
&= \frac{\pi e^{-3} e^{2i}}{3} \\
&= \frac{\pi}{3e^3}(\cos 2 + i \sin 2).
\end{aligned}
$$

(iii) By Jordan's Lemma, with $a = 1$ and $f(z) = 1/(z^2 - 4z + 13)$,

$$\int_{C_R^+} \frac{e^{iz}}{z^2 - 4z + 13} \, dz \to 0 \text{ as } R \to \infty.$$

(iv) Now we shall calculate the integrals along the two parts of the contour. Along γ_R we get

$$\int_{\gamma_R} \frac{e^{iz}}{z^2 - 4z + 13}\, dz = \int_{-R}^{R} \frac{e^{ix}}{x^2 - 4x + 13}\, dx$$

$$= \int_{-R}^{R} \frac{\cos x}{x^2 - 4x + 13}\, dx + i \int_{-R}^{R} \frac{\sin x}{x^2 - 4x + 13}\, dx.$$

Therefore,

$$\lim_{R \to \infty} \int_{\gamma} f(z)\, dz = \lim_{R \to \infty} \int_{\gamma_R} f(z)\, dz + \lim_{R \to \infty} \int_{C_R^+} f(z)\, dz$$

$$= \lim_{R \to \infty} \left\{ \int_{-R}^{R} \frac{\cos x}{x^2 - 4x + 13}\, dx + i \int_{-R}^{R} \frac{\sin x}{x^2 - 4x + 13}\, dx \right\} + 0$$

$$\frac{\pi}{3e^3}(\cos 2 + i \sin 2) = \int_{-\infty}^{\infty} \frac{\cos x}{x^2 - 4x + 13}\, dx + i\, \mathrm{PV} \int_{-\infty}^{\infty} \frac{\sin x}{x^2 - 4x + 13}\, dx.$$

Equating real parts we get

$$\int_{-\infty}^{\infty} \frac{\cos x}{x^2 - 4x + 13}\, dx = \frac{\pi \cos 2}{3e^3}.$$

Can you imagine calculating this integral using the standard methods of real analysis?

Remark 19.10

Note that using the imaginary parts we get

$$\mathrm{PV} \int_{-\infty}^{\infty} \frac{\sin x}{x^2 - 4x + 13}\, dx = \frac{\pi \sin 2}{3e^3}.$$

Thus, we have gained more information than we were looking for. What a method!

Jordan's Lemma (Strong Version)

Although the earlier version of Jordan's Lemma can be used in most circumstances there is a slightly stronger version that is useful and hence is worth knowing.

Lemma 19.11 (Jordan's Lemma(Strong Version))

Suppose that f is a complex function which is analytic at z for $\mathrm{Re}(z) \geq 0$ except for a finite number of points. Suppose also that

$$M_R = \sup_{z \in (C_R^+)^*} |f(z)| \to 0 \text{ as } R \to \infty.$$

Then, for all $a > 0$,

$$\int_{C_R^+} e^{iaz} f(z) \, dz \to 0 \text{ as } R \to \infty.$$

Proof. Let

$$I_R = \int_{C_R^+} e^{iaz} f(z) \, dz.$$

Then

$$|I_R| = \left| \int_0^\pi e^{iaR(\cos t + i \sin t)} f(Re^{it}) \left(iRe^{it} \right) \, dt \right|$$

$$\leq \int_0^\pi \left| e^{iaR(\cos t + i \sin t)} f(Re^{it}) \left(iRe^{it} \right) \right| \, dt$$

$$= R \int_0^\pi e^{-aR \sin t} \left| f(Re^{it}) \right| \, dt$$

$$\leq R \int_0^\pi e^{-aR \sin t} \sup_{z \in (C_R^+)^*} |f(z)| \, dt$$

$$= R M_R \int_0^\pi e^{-aR \sin t} \, dt$$

But since sin is symmetric about $\pi/2$ we get

$$|I_R| \leq R M_R \times 2 \int_0^{\pi/2} e^{-aR \sin t} \, dt$$

Now note that $\sin t \geq \dfrac{2t}{\pi}$ for $0 \leq t \leq \pi/2$. (The rigorous derivation of this result is left as an exercise but drawing a graph is fairly convincing.) So,

$$|I_R| \leq 2 R M_R \int_0^{\pi/2} e^{-2aRt/\pi} \, dt$$

$$= 2 R M_R \left(\frac{\pi}{2aR} \left(1 - e^{-aR} \right) \right)$$

$$\leq \frac{\pi M_R}{a}$$

$$\to 0 \text{ as } R \to \infty.$$

$\square$

Exercise 19.12

The statement is not true in the case of $a = 0$. Give a counter example.

Example 19.13
Evaluate the integral $\int_0^\infty \frac{x^3 \sin x}{(x^2+1)^2} \, dx$.

First note that as the integrand is an even function we have

$$\int_0^\infty \frac{x^3 \sin x}{(x^2+1)^2} \, dx = \frac{1}{2} \, \mathrm{PV} \int_{-\infty}^\infty \frac{x^3 \sin x}{(x^2+1)^2} \, dx.$$

As usual we take complexify with e^{iz} substituted for $\sin x$ so we have

$$f(z) = \frac{z^3 e^{iz}}{(z^2+1)^2}.$$

This has poles at $z = \pm i$. Only i is in the upper half plane and the residue has been calculated in Example 17.8(iv):

$$\mathrm{res}(f, i) = \frac{1}{4e}.$$

Let γ, γ_R and C_R^+ denote our usual contours. By the strong version of Jordan's Lemma

$$\lim_{R \to \infty} \int_{C_R^+} f(z) \, dz = 0.$$

We have

$$\int_\gamma f = \int_{\gamma_R} f + \int_{C_R^+} f$$

$$2\pi i \frac{1}{4e} = \lim_{R \to \infty} \int_{-R}^R \frac{\cos x + i \sin x}{(x^2+1)^2} \, dx$$

$$\frac{\pi}{2e} = \mathrm{PV} \int_{-\infty}^\infty \frac{x^3 \sin x}{(x^2+1)^2} \, dx$$

$$\frac{\pi}{4e} = \int_0^\infty \frac{x^3 \sin x}{(x^2+1)^2} \, dx.$$

Multiple versions of Jordan's Lemma exist. Here is another.

Lemma 19.14 (Jordan's Lemma (Weak version II))
Suppose that f is a complex function which is analytic at z for $\mathrm{Re}(z) \geq 0$ except for a finite number of points. Suppose also that there exists real K and R' such that

$$|f(z)| \leq \frac{K}{R^2} \text{ for all } R > R'.$$

Then, for all $a \geq 0$,

$$\int_{C_R^+} e^{iaz} f(z) \, dz \to 0 \text{ as } R \to \infty.$$

Proof. The proof is mostly an application of the Estimation Lemma and is left as an exercise. $\square$

Remark 19.15

Suppose that p and q are polynomials with $\deg(p) \leq \deg(q) - 2$. Then there exist K and R' such that $|f(z)| \leq KR^{-2}$ for all $R > R'$. (Exercise.) Hence the weak version II of Jordan's Lemma applies in this case.

Exercises

Exercises 19.16

(i) Find the following.

(a) $\displaystyle \int_{-\infty}^{\infty} \frac{1}{x^2 - 2x + 5}\, dx,$

(b) $\displaystyle \int_{-\infty}^{\infty} \frac{1}{(x^2 + 1)(x^2 + 4)}\, dx,$

(c) $\displaystyle \int_{-\infty}^{\infty} \frac{1}{(x^2 + a^2)(x^2 + b^2)}\, dx,$ where $a > 0, b > 0, a \neq b,$

(d) $\displaystyle \int_{-\infty}^{\infty} \frac{\cos x}{x^4 + 2x^2 + 1}\, dx,$

(e) $\displaystyle \int_{-\infty}^{\infty} \frac{\sin x}{x(x^2 + 9)}\, dx,$

(f) $\displaystyle \int_{-\infty}^{\infty} \frac{x^2}{x^4 - 2x^2 + 5}\, dx.$

(ii) Find formula for the following where $b^2 < 4ac$, $a, b, c \in \mathbb{R}$:

(a) $\displaystyle \int_{-\infty}^{\infty} \frac{1}{ax^2 + bx + c}\, dx,$

(b) $\displaystyle \int_{-\infty}^{\infty} \frac{1}{(ax^2 + bx + c)^2}\, dx.$

(iii) Find

(a) $\displaystyle \int_{0}^{\infty} \frac{x}{1 + x^4}\, dx,$ (b) $\displaystyle \int_{0}^{\infty} \frac{1}{(x^2 + 1)^4}\, dx.$

(iv) Find, for $a > 0$,

$$\int_{0}^{\infty} \frac{\cos x}{x^2 + a^2}\, dx.$$

(v) Show that
$$\int_0^\infty \frac{\sin 2x}{x(1+x^2)} \, dx = \frac{\pi}{2} \left(1 - e^{-2}\right).$$

(vi) Calculate
$$\int_{-\infty}^\infty \frac{\cos x + x \sin x}{(x^2+1)^2} \, dx.$$

(vii) (a) Use the slice of pie contour from Exercise 11.12(vi) to find the integral from Example 19.7, $\int_0^\infty \frac{1}{1+x^{16}} \, dx$. [Hint: Relate the integral on the positive real axis to the one returning to the origin.]

Which method do you prefer for this integral?

(b) As seen in Exercise 19.8 we can generalize Example 19.7. Calculate
$$\int_0^\infty \frac{1}{1+x^n} \, dx$$

for all $1 \neq n \in \mathbb{N}$.

(viii) Consider the bump contour, $\Gamma = \gamma_R + \gamma^- - \gamma_\varepsilon + \gamma^+$ from Exercise 5.18(vi) where $0 < \varepsilon < R$.

(a) Calculate $\int_\Gamma \frac{e^{iz}}{z} \, dz$.

(b) Show that $\int_{\gamma_\varepsilon} \frac{e^{iz}}{z} \, dz = -\pi i$.

(c) Find $\int_\varepsilon^R \frac{\sin x}{x} \, dx$ in terms of integrals of e^{iz}/z on the real line.

(d) Hence show that $\int_0^\infty \frac{\sin x}{x} \, dx = \frac{\pi}{2}$.

This is an example of a **Dirichlet integral**.

(ix) Use the bump contour from the previous exercise and $(1 - e^{2iz})/z^2$ to find
$$\int_{-\infty}^\infty \frac{\sin^2 x}{x^2} \, dx.$$

(x) By taking bumps at appropriate positions show that, for $a > 0$,
$$\int_{-\infty}^\infty \frac{\cos x}{a^2 - x^2} \, dx = \frac{\pi \sin a}{a}.$$

(xi) State and prove a theorem that encapsulates the semi-circular method from this chapter.

Many possible answers. For example: Let $f : \mathbb{R} \to \mathbb{R}$ be such that there exists a complex function F is with $F(z) = f(z)$ for all $z \in \mathbb{R}$ and F is differentiable except at a finite number of points. If $\sup\limits_{|z|=R, \mathrm{Im}(z) \geq 0} |F(z)| \to 0$ as $R \to \infty$, then for all $a > 0$,

$$\mathrm{PV} \int_{-\infty}^{\infty} e^{iaz} f(x) \, dx = 2\pi i \times \text{ sum of residues of } F \text{ in upper half-plane.}$$

(xii) Prove Lemma 19.14 (Jordan's Lemma Weak version II).

(xiii) Prove that $\sin t \geq \dfrac{2t}{\pi}$ for $0 \leq t \leq \pi/2$.

(xiv) Give a counter example to the statement of Jordan's Lemma (Strong version) in the case $a = 0$.

(xv) Show that the weak version of Jordan's Lemma can be deduced from the strong version. Furthermore, if we replace $\deg(p) \leq \deg(q) - 2$ with $\deg(p) \leq \deg(q) - 1$ in the weak version, what other condition do we need for the resulting statement to be true?

Summary

___

❑ (i) Define $\gamma = \gamma_R + C_R^+$ by

$$\gamma_R(t) = t \text{ for } -R \leq t \leq R, \text{ and } C_R^+(t) = Re^{it} \text{ for } 0 \leq t \leq \pi.$$

Take R large enough to include the poles of f that lie in the upper half-plane.

(ii) Calculate $\int_\gamma f(z) \, dz$ using Cauchy's Residue Theorem.

(iii) Show that $\int_{C_R^+} f(z) \, dz \to 0$ as $R \to \infty$, using Jordan's Lemma.

(iv) Relate $\int_{-\infty}^{\infty} f(x) \, dx$ to $\int_\gamma f(z) \, dz$.

❑ The method does not always calculate the integral we want – it calculates the principal value of the integral.

❑ Many other contours can be used.

CHAPTER 20

Real Integrals of Functions of Sines and Cosines

Let us consider the nasty integral

$$\int_0^{2\pi} e^{\cos\theta} \cos(n\theta - \sin\theta)\, d\theta.$$

from Chapter 5. (Note the similarity to the Bessel function J_n of Exercise 15.11(iv) evaluated at 1.) This is very hard to evaluate with the methods of integration taught in a standard calculus course.

If we consider contour integration, then is there some function and contour such that $\int_\gamma f$ gives the above? Let $\gamma(\theta) = e^{i\theta}$, for $0 \le \theta \le 2\pi$ and $f(z) = \dfrac{e^z}{z^{n+1}}$. Admittedly this choice of f appears to come from nowhere and it is not obvious that such a choice is the right one for the nasty integral. It will however, take us where we want to go.

With our knowledge of poles and residues we can calculate that

$$\int_\gamma \frac{e^z}{z^{n+1}}\, dz = 2\pi i \ \mathrm{res}(e^z/z^{n+1}, 0) = \frac{2\pi i}{n!}.$$

(This really is not too difficult, use Method D).

Hence,

$$\int_\gamma \frac{e^z}{z^{n+1}} \, dz = \int_0^{2\pi} \frac{e^{e^{i\theta}}}{e^{i\theta(n+1)}} i e^{i\theta} \, d\theta$$

$$\frac{2\pi i}{n!} = i \int_0^{2\pi} e^{e^{i\theta}} e^{-i\theta n} \, d\theta$$

$$\frac{2\pi}{n!} = \int_0^{2\pi} e^{e^{i\theta} - in\theta} \, d\theta$$

$$= \int_0^{2\pi} e^{\cos\theta + i\sin\theta - in\theta} \, d\theta$$

$$= \int_0^{2\pi} e^{\cos\theta} e^{i(\sin\theta - n\theta)} \, d\theta$$

$$= \int_0^{2\pi} e^{\cos\theta} \left(\cos(\sin\theta - n\theta) + i \sin((\sin\theta - n\theta)) \right) \, d\theta.$$

By equating real and imaginary parts we see that

$$\int_0^{2\pi} e^{\cos\theta} \cos(\sin\theta - n\theta) \, d\theta = \frac{2\pi}{n!}$$

and because cos is an even function we deduce that

$$\int_0^{2\pi} e^{\cos\theta} \cos(n\theta - \sin\theta) \, d\theta = \frac{2\pi}{n!}$$

Rather spectacular, wouldn't you agree?

Integrals of the form $\int_0^{2\pi} f(\cos\theta, \sin\theta) \, d\theta$

In this section we will deal with functions of sine and cosine (that usually are not as complicated as that last example!).

Theorem 20.1
Let $\gamma(t) = e^{it}$, for $0 \le t \le 2\pi$. Then

$$\int_0^{2\pi} f(\cos\theta, \sin\theta) \, d\theta = -i \int_\gamma f\left(\frac{z + z^{-1}}{2}, \frac{z - z^{-1}}{2i} \right) z^{-1} \, dz.$$

Proof. Let $z = e^{i\theta} = \gamma(\theta)$ for $0 \le \theta \le 2\pi$. Then $\cos\theta = \dfrac{z + z^{-1}}{2}$ and $\sin\theta = \dfrac{z - z^{-1}}{2i}$. Also, we have $\gamma'(\theta) = ie^{i\theta}$.

Thus,

$$-i \int_\gamma f\left(\frac{z+z^{-1}}{2}, \frac{z-z^{-1}}{2i}\right) z^{-1}\, dz = -i \int_0^{2\pi} f(\cos\theta, \sin\theta) e^{-i\theta} i e^{i\theta}\, d\theta$$

$$= \int_0^{2\pi} f(\cos\theta, \sin\theta)\, d\theta.$$

□

Like many mathematical formulae, it is easier to remember if you understand where all the terms come from.

Example 20.2
Evaluate the real integral $\int_0^{2\pi} \dfrac{1}{13+12\cos t}\, dt$.

Solution: Applying the theorem we get, for $\gamma(t) = e^{it}$, $0 \le t \le 2\pi$,

$$\int_0^{2\pi} \frac{1}{13+12\cos t}\, dt = -i \int_\gamma \frac{1}{z\,(13+6\,(z+1/z))}\, dz$$

$$= -i \int_\gamma \frac{1}{6z^2+13z+6}\, dz.$$

Let $g(z) = \dfrac{1}{6z^2+13z+6}$. This has poles at $-\frac{3}{2}$ and $-\frac{2}{3}$ since $6z^2+13z+6 = (3z+2)(2z+3)$. Of these, only $-\frac{2}{3}$ lies within the unit circle given by γ. Hence, we calculate the residue for this pole:

$$\operatorname{res}\left(g, -\frac{2}{3}\right) = \frac{1}{12z+13}\Big|_{z=-2/3} = \frac{1}{12\left(-\frac{2}{3}+13\right)+13} = \frac{1}{5}.$$

Therefore, by Cauchy's Residue Theorem,

$$\int_0^{2\pi} \frac{1}{13+12\cos t}\, dt = -i \times 2\pi i \times \frac{1}{5} = \frac{2\pi}{5}.$$

Example 20.3
Evaluate the integral $\int_0^{2\pi} \sin^6 x\, dx$.

Solution: As before let γ be the contour describing the standard unit circle contour. We have

$$\int_0^{2\pi} \sin^6 x \, dx = -i \int_\gamma \left(\frac{z - 1/z}{2i} \right)^6 \frac{1}{z} \, dz$$

$$= -i \int_\gamma \frac{1}{(-64)} (-20 + \text{other terms}) \frac{1}{z} \, dz, \text{ by the binomial theorem,}$$

$$= -i \times 2\pi i \left(\frac{20}{64} \right), \text{ by Cauchy's Residue Formula,}$$

$$= \frac{5\pi}{8}.$$

Note that we did not need to calculate all the terms in the binomial expansion of $(z - 1/z)^6$. Due to the $1/z$ factor in the integrand we could anticipate that to calculate the residue we needed only the constant term from the binomial expansion.

Remark 20.4
We can change the limits of integration by letting $\gamma(t) = e^{it}$ where $a \le t \le a + 2\pi$ and $a \in \mathbb{R}$. So we get integrals of the form $\int_a^{a+2\pi} f(\cos \theta, \sin \theta) \, d\theta$. A common example is $a = -\pi$, so we have $\int_{-\pi}^{\pi}$.

Since in Theorem 20.1 the integral is calculated using residues and these only depend on the interior of γ and not its start and end point, we have

$$\int_a^{a+2\pi} f(\cos \theta, \sin \theta) \, d\theta = \int_0^{2\pi} f(\cos \theta, \sin \theta) \, d\theta.$$

Example 20.5
We have

$$\int_{-\pi}^{\pi} \frac{1}{4 + \sin t} \, dt = \int_0^{2\pi} \frac{1}{4 + \sin t} \, dt.$$

This can now be calculated with the usual method. (In this case the integral is $2\pi/\sqrt{15}$).

Using De Moivre's Theorem

From De Moivre's Theorem we have

$$e^{in\theta} = (\cos \theta + i \sin \theta)^n = \cos n\theta + i \sin \theta$$

and so we can deduce that

$$\cos n\theta = \frac{1}{2}\left(e^{in\theta} + e^{-in\theta}\right) = \frac{1}{2}\left(z^n + z^{-n}\right),$$

$$\sin n\theta = \frac{1}{2i}\left(e^{in\theta} - e^{-in\theta}\right) = \frac{1}{2i}\left(z^n - z^{-n}\right).$$

for $z = e^{i\theta}$.

Example 20.6

Find

$$\int_0^{2\pi} \frac{\cos 2\theta}{(5 - 3\cos\theta)^2}\, d\theta.$$

Solution: For $\gamma(t) = e^{it}$, $0 \le t \le 2\pi$, the integral is equivalent to

$$-i \int_\gamma \frac{\frac{1}{2}\left(z^2 + z^{-2}\right)}{\left(5 - \frac{3}{2}\left(z + z^{-1}\right)\right)^2} \frac{1}{z}\, dz$$

$$= -\frac{i}{2} \int_\gamma \frac{\left(z^2 + z^{-2}\right) z}{\frac{1}{4}\left(10 - 3\left(z + z^{-1}\right)\right)^2 z^2}\, dz$$

$$= -2i \int_\gamma \frac{z^3 + z^{-1}}{\left(10z - 3z^2 - 3\right)^2}\, dz$$

$$= -2i \int_\gamma \frac{z^4 + 1}{z\left(3z - 1\right)^2 \left(z - 3\right)^2}\, dz.$$

The integrand has poles at 0, 1/3 and 3. Only the first two are within the unit circle so we calculate the residues of the integrand at these points. After some calculation, we find that these are 1/9 and $-115/1152$ respectively. Hence,

$$\int_0^{2\pi} \frac{\cos 2\theta}{(5 - 3\cos\theta)^2}\, d\theta = -2i \times 2\pi i \times \left(\frac{1}{9} - \frac{115}{1152}\right) = \frac{13\pi}{288}.$$

Exercises

Exercises 20.7

(i) Calculate the following

(a) $\displaystyle\int_0^{2\pi} \sin^2 t \, dt$

(b) $\displaystyle\int_{-\pi}^{\pi} \frac{1}{1 + 12\cos^2 t} \, dt$

(c) $\displaystyle\int_0^{2\pi} \frac{\cos t}{17 - 8\cos t} \, dt$

(d) $\displaystyle\int_0^{2\pi} \cos^{2n}(\theta) \, d\theta.$

(e) $\displaystyle\int_0^{\pi} \frac{1}{10 + 8\cos(\theta)} \, d\theta,$

(f) $\displaystyle\int_0^{\pi} \frac{1}{(2 + \cos\theta)^2} \, d\theta,$

(g) $\displaystyle\int_{-\pi}^{\pi} \frac{1}{4 + \sin^4 t} \, dt,$

(h) $\displaystyle\int_0^{2\pi} \cos^4 t + 3\sin^4 t \, dt,$

(i) $\displaystyle\int_0^{\pi/2} \frac{1}{a + \sin^2 t} \, dt, \ (a > 0),$

(j) $\displaystyle\int_0^{2\pi} \cos^4 t + 3\sin^4 t \, dt.$

(ii) Show that

$$\int_0^{2\pi} \frac{1}{a + b\cos^2\theta} \, d\theta = \frac{2\pi}{\sqrt{a}\sqrt{a+b}}, \qquad \text{for } 0 < b < a.$$

(iii) Find

(a) $\displaystyle\int_0^{2\pi} \frac{1}{1 - 2a\cos\theta + a^2} \, d\theta, \qquad \text{for } 0 \neq |a| \neq 1,$

(b) $\displaystyle\int_0^{\pi} \frac{\cos 2\theta}{1 - 2a\cos\theta + a^2} \, d\theta, \qquad \text{for } 0 < |a| < 1.$

(iv) Let f be analytic on $|z| < R$ for some $R > 1$.

(a) Show that

$$\int_0^{2\pi} f\left(e^{it}\right) \cos^2\left(\frac{t}{2}\right) dt = \pi \left(f(0) + \frac{f'(0)}{2}\right).$$

(b) Find and prove a similar statement with cos replaced by sin.

(c) From (a) show that, for $n \in \mathbb{N}$ and $n > 1$,

$$\int_0^{2\pi} \cos(nt) \cos^2\left(\frac{t}{2}\right) dt = 0.$$

(v) Show that, for $0 \neq a \in \mathbb{R}$,

$$\int_0^{2\pi} \cot(t + ia)\, dt = -2\pi i \operatorname{sign}(a).$$

(vi) Let $f(x) = \exp\left(\frac{x}{2}\left(z + \frac{1}{z}\right)\right)$ and let $J_n(x) = \frac{1}{2\pi} \int_0^{2\pi} \cos(n\theta - x \sin\theta)\, d\theta$ be the nth Bessel function.

(a) Prove that

$$f(x) = \sum_{-\infty}^{\infty} J_n(x) z^n.$$

(b) Show that

$$J_n(x) = (-1)^n J_{-n}(x)$$

and hence deduce

$$f(x) = J_0(x) + \sum_{n=1}^{\infty} J_n(x)\left(z^n + z^{-n}\right).$$

(c) Prove that

$$J_n(x) = \left(\frac{x}{2}\right)^n \sum_{k=0}^{\infty} \frac{(-1)^k}{k!(n+k)!} \left(\frac{x}{2}\right)^{2k}.$$

Note that this shows that Bessel functions are analytic.

Summary

❑ Let $\gamma(t) = e^{it}$, for $0 \leq t \leq 2\pi$. Then

$$\int_0^{2\pi} f(\cos\theta, \sin\theta)\, d\theta = -i \int_\gamma f\left(\frac{z + z^{-1}}{2}, \frac{z - z^{-1}}{2i}\right) z^{-1}\, dz.$$

CHAPTER 21

Summation of Series

In courses introducing convergence of series, the series

$$\sum_{n=1}^{\infty} \frac{1}{n^2}$$

is shown to converge and is often used, via the comparison test, to show other series converge. Students are justified in asking 'But, what does it actually converge *to*?' We are now in a position to show that

$$\sum_{n=1}^{\infty} \frac{1}{n^2} = \frac{\pi^2}{6}.$$

This was originally discovered by Euler but, like much of his work, his level of rigour is not acceptable today.

The result follows from the following theorem.

Theorem 21.1 (First summation theorem)
Suppose that

(i) *$f : \mathbb{C} \to \mathbb{C}$ is analytic except for poles at $p_1, \ldots, p_m$,*

(ii) *$f(n) \neq 0$ for $n \in \mathbb{Z} \backslash \{p_1, \ldots, p_m\}$,*

(iii) *there exists constants K and R such that*

$$|f(z)| \leq \frac{K}{|z|^2} \text{ for all } |z| \geq R.$$

Then, the series

$$\sum_{\substack{n=-\infty \\ n \neq p_j}}^{\infty} f(n)$$

converges and

$$\sum_{\substack{n=-\infty \\ n \neq p_j}}^{\infty} f(n) = -\sum_{j=1}^{m} \operatorname{res}(\pi f(z) \cot(\pi z), p_j).$$

Let's see this in action before proving it.

Example 21.2

To find the value of $\displaystyle\sum_{n=1}^{\infty} \frac{1}{n^2}$ we apply the theorem with $f(z) = \dfrac{1}{z^2}$. This function is analytic except for a pole at 0. The function obviously satisfies conditions (ii) and (iii). Therefore,

$$
\begin{aligned}
\sum_{\substack{n=-\infty \\ n \neq 0}}^{\infty} \frac{1}{n^2} &= -\operatorname{res}\left(\frac{1}{z^2}\pi \cot(\pi z), 0\right) \\
&= -\left(-\frac{\pi^2}{3}\right) \quad \text{by Examples 17.8(i)}, \\
&= \frac{\pi^2}{3}.
\end{aligned}
$$

So

$$\sum_{\substack{n=-\infty \\ n \neq 0}}^{\infty} \frac{1}{n^2} = \sum_{n=-\infty}^{n=-1} \frac{1}{n^2} + \sum_{n=1}^{\infty} \frac{1}{n^2} = 2\sum_{n=1}^{\infty} \frac{1}{n^2},$$

and

$$\sum_{n=1}^{\infty} \frac{1}{n^2} = \frac{1}{2}\frac{\pi^2}{3} = \frac{\pi^2}{6}.$$

To prove the theorem we first need a lemma.

Lemma 21.3

Let γ_N be a path that describes the boundary of the square with corners at $\pm(N + \frac{1}{2}) \pm i(N + \frac{1}{2})$. Then, for all $z \in \gamma_N^*$,

$$|\cot \pi z| \leq \sqrt{2}.$$

Proof. In the same way as Exercise 3.16(xiii) we can show that

$$\left|\sin^2(a+ib)\right| = \sin^2 a + \sinh b^2.$$

If $z = \pm(N + \frac{1}{2}) + iy$, i.e., z lies on a vertical part of γ_N^*, then

$$\left|\sin^2 \pi z\right| = \sin^2(N + \frac{1}{2})\pi + \sinh^2 \pi y \geq 1 + 0 = 1.$$

If $z = x \pm i(N + \frac{1}{2})$, i.e., z lies on a horizontal part of γ_N^*, then

$$\left|\sin^2 \pi z\right| = \sin^2 \pi x + \sinh^2(N + \frac{1}{2})\pi \geq 0 + \sinh^2(\pi/2) \geq 1.$$

So, $\left|\sin^2 \pi z\right| \geq 1$ on γ_N^*, hence $\left|\operatorname{cosec}^2 \pi z\right| \leq 1$ and

$$\left|\cot^2 \pi z\right| = \left|\operatorname{cosec}^2 \pi z - 1\right| \leq \left|\operatorname{cosec}^2 \pi z\right| + 1 \leq 2,$$

i.e., $\left|\cot \pi z\right| \leq \sqrt{2}$.   □

Proof (of first summation theorem). If $|n| \geq R$, then for $n \neq p_j$,

$$|f(n)| \leq \frac{K}{n^2}$$

so

$$\sum_{\substack{n=R \\ n\neq p_j}}^{\infty} |f(n)|$$

converges by the comparison test and hence, using Theorem 2.24, the series

$$\sum_{\substack{n=0 \\ n\neq p_j}}^{\infty} f(n)$$

converges. The sum of terms with negative index converges in a similar manner so the series of the theorem converges.

Let γ_N be the contour describing a s square in Lemma 21.3, and assume that we have $N > \max\{|p_1|, \ldots, |p_m|, R\}$. We are going to consider the integral

$$\int_{\gamma_N} f(z)\pi \cot \pi z \, dz$$

as $N \to \infty$. If $z \in \gamma_N^*$, then $|z| \geq N$, thus by the assumption on f,

$$|f(z)| \leq \frac{K}{N^2}.$$

We know from Lemma 21.3 that $|\cot \pi z| \leq \sqrt{2}$ and it is easy to show that $L(\gamma_N) = 8(N + \frac{1}{2})$, so by the Estimation Lemma

$$\left| \int_{\gamma_N} f(z)\pi \cot \pi z \, dz \right| \leq \frac{K}{N^2} \times \pi \times \sqrt{2} \times 8 \left(N + \frac{1}{2} \right).$$

By letting $N \to \infty$ we see that the integral tends to 0.

The function

$$f(z)\pi \cot \pi z = f(z)\pi \frac{\cos \pi z}{\sin \pi z}$$

has poles at $p_1, \ldots, p_m$ and at zeroes of $\sin \pi z$, i.e., when $z \in \mathbb{Z}$. If $n \in \mathbb{Z}$ and $n \neq p_j$, then, by assumption on f and Method B,

$$\text{res}\,(f(z)\pi \cot \pi z, n) = \text{res}\left(\frac{f(z)\pi \cos \pi z}{\sin \pi z}, n \right) = \frac{f(n)\pi \cos \pi n}{\pi \cos \pi n} = f(n).$$

From Cauchy's Residue Theorem we have

$$\int_{\gamma_N} f(z)\pi \cot \pi z \, dz = 2\pi i \left\{ \sum_{j=1}^{m} \text{res}(f(z)\pi \cot \pi z, p_j) + \sum_{\substack{n=-N \\ n \neq p_j}}^{N} \text{res}(f(z)\pi \cot \pi z, n) \right\}.$$

Therefore,

$$\begin{aligned}
0 &= \lim_{N \to \infty} \int_{\gamma_N} f(z)\pi \cot \pi z \, dz \\
&= \lim_{N \to \infty} 2\pi i \left\{ \sum_{j=1}^{m} \text{res}(f(z)\pi \cot \pi z, p_j) + \sum_{\substack{n=-N \\ n \neq p_j}}^{N} f(n) \right\} \\
&= 2\pi i \left\{ \sum_{j=1}^{m} \text{res}(f(z)\pi \cot \pi z, p_j) + \sum_{\substack{n=-\infty \\ n \neq p_j}}^{\infty} f(n) \right\}.
\end{aligned}$$

$\square$

Exercise 21.4
Find the value of $\displaystyle\sum_{n=1}^{\infty} \frac{1}{n^4}$.

It is possible to calculate $\displaystyle\sum_{n=1}^{\infty} \frac{1}{n^k}$ for any even natural number k but, perhaps surprisingly, for odd k no one knows what the sum is in a simple form. In particular, it is an open problem to find $\sum_{n=1}^{\infty} \dfrac{1}{n^3}$. This number is called Apéry's constant.

Example 21.5

We will now show that

$$\sum_{n=1}^{\infty} \frac{1}{n^2 + a^2} = \frac{\pi \coth \pi a}{2a} - \frac{1}{2a^2}$$

for all $a \in \mathbb{C}$ such that $a \neq im$ for any integer m.

Let $f(z) = \dfrac{1}{z^2 + a^2}$, then $f(n) \neq 0$, $|f(z)| \leq 1/|z|^2$ for $|z| > |a|$. The poles are at $z = \pm ia$. We need to calculate $\mathrm{res}(\pi f(z) \cot \pi z, \pm ia)$. Since $\cot \pi(\pm ia) \neq 0$ we let $p(z) = \pi \cot \pi z$ and $q(z) = z^2 + a^2$ in Method B to get

$$
\begin{aligned}
\mathrm{res}\left(\pi \frac{1}{z^2 + a^2} \cot \pi z, \pm ia\right) &= \frac{\pi \cot \pi(\pm ia)}{2(\pm ia)} \\
&= -\frac{\pm \pi i \coth \pi a}{\pm 2ia}, \quad \text{as } \cot(\lambda i) = -i \cot(\lambda), \\
&= -\frac{\pi \coth \pi a}{2a}.
\end{aligned}
$$

Thus the residues at $-ia$ and ia are the same.

From the summation theorem (taking p_1 and p_2 as our poles $\pm ia$),

$$\sum_{n=-\infty}^{\infty} f(n) = -\sum_{j=1}^{2} \mathrm{res}(\pi f(z) \cot \pi z, p_j)$$

$$\frac{1}{a^2} + 2\sum_{n=0}^{\infty} \frac{1}{n^2 + a^2} = -\left(-2\frac{\pi \coth \pi a}{2a}\right)$$

$$\sum_{n=1}^{\infty} \frac{1}{n^2 + a^2} = \frac{\pi \coth \pi a}{2a} - \frac{1}{2a^2}.$$

Exercise 21.6

By replacing a in the formula with ia show that

$$\sum_{n=1}^{\infty} \frac{1}{n^2 - a^2} = \frac{1}{2a^2} - \frac{\pi \cot \pi a}{2a}$$

for all $a \notin \mathbb{Z}$.

Example 21.7

Example 21.5 can be used to give new expression for $\coth z$. Let $z = \pi a$, then we get

$$\frac{\pi \coth z}{2(z/\pi)} = \sum_{n=1}^{\infty} \frac{1}{n^2 + (z/\pi)^2} + \frac{1}{2(z/\pi)^2}$$

$$\frac{\pi^2 \coth z}{2z} = \sum_{n=1}^{\infty} \frac{\pi^2}{\pi^2 n^2 + z^2} + \frac{\pi^2}{2z^2}$$

$$\coth z = \frac{1}{z} + 2z \sum_{n=1}^{\infty} \frac{1}{z^2 + \pi^2 n^2}.$$

This holds for all $z \neq i\pi m$ for any integer m.

Exercise 21.8

Show that

$$\cot z = \frac{1}{z} + 2z \sum_{n=1}^{\infty} \frac{1}{z^2 - \pi^2 n^2}$$

for all $z \neq \pi m$, $m \in \mathbb{Z}$.

We can push a little bit further the type of series we can sum as the following shows.

Example 21.9

We will show that

$$\sum_{n=1}^{\infty} \frac{1}{16n^2 - 1} = \frac{1}{2} - \frac{\pi}{8}.$$

We have

$$\sum_{n=1}^{\infty} \frac{1}{16n^2 - 1} = \sum_{n=1}^{\infty} \frac{1}{16(n^2 - (1/4)^2)}$$

$$= \frac{1}{16} \sum_{n=1}^{\infty} \frac{1}{n^2 - (1/4)^2}$$

$$= \frac{1}{16} \left(\frac{1}{2(1/4)^2} - \frac{\pi \cot(\pi/4)}{2(1/4)} \right)$$

$$= \frac{1}{16} \left(\frac{16}{2} - 2\pi \right)$$

$$= \frac{1}{2} - \frac{\pi}{8}.$$

It is important to note that complex analysis is allowing us to evaluate sums that might otherwise be difficult to calculate.

Alternating series

We can also find the values of alternating series.

Theorem 21.10 (Second summation theorem)
With the assumptions of Theorem 21.1 the following series converges to the given value

$$\sum_{\substack{n=-\infty \\ n \neq p_j}}^{\infty} (-1)^n f(n) = -\sum_{j=1}^{m} \operatorname{res}\left(\frac{\pi f(z)}{\sin(\pi z)}, p_j\right),$$

Proof. The proof is similar to the proof of the first summation theorem. Convergence again comes from application of the comparison test. For the estimation lemma part we deduce from the proof of Lemma 21.3 that $|\sin(\pi z)| \geq 1$ for $z \in \gamma_N^*$. For the residues it is easy to calculate that they are $(-1)^n f(n)$ for $n \in \mathbb{Z}$ and $n \neq p_j$.

$\square$

Example 21.11
We shall show

$$\sum_{n=1}^{\infty} \frac{(-1)^{n+1}}{n^2} = \frac{\pi^2}{12}.$$

Consider again the function $f(z) = 1/z^2$. From the theorem we have that

$$\sum_{\substack{n=-\infty \\ n \neq 0}}^{\infty} \frac{(-1)^n}{n^2} = -\operatorname{res}\left(\frac{\pi}{z^2 \sin(\pi z)}, 0\right).$$

The residue is $\dfrac{\pi^2}{6}$ since by Example 15.6(i)

$$\frac{1}{z^2 \sin \pi z} = \frac{1}{z^2}\left(\frac{1}{\pi z} + \frac{\pi z}{6} + \frac{7\pi^3 z^3}{360} + \cdots\right).$$

Hence,

$$2\sum_{n=1}^{\infty} \frac{(-1)^n}{n^2} = -\frac{\pi^2}{6}$$

i.e.,

$$\sum_{n=1}^{\infty} \frac{(-1)^{n+1}}{n^2} = \frac{\pi^2}{12}.$$

Example 21.12
Let us evaluate

$$\sum_{n=0}^{\infty} \frac{(-1)^n}{(2n+1)^3}.$$

Let $f(z) = \dfrac{1}{(2z+1)^3}$. The function $\pi f(z)/\sin(\pi z)$ has a pole of order 3 at $z = -1/2$. Thus, using Method C we get

$$\mathrm{res}\left(\frac{\pi}{(2z+1)^3 \sin(\pi z)}, -\frac{1}{2}\right)$$

$$= \frac{1}{(3-1)!} \lim_{z \to -1/2} \frac{d^2}{dz^2}\left((z+1/2)^3 \frac{\pi}{8(z+1/2)^3 \sin(\pi z)}\right)$$

$$= \frac{\pi}{16} \lim_{z \to -1/2} \frac{d^2}{dz^2}\left(\frac{1}{\sin(\pi z)}\right)$$

$$= \frac{\pi}{16} \lim_{z \to -1/2} \frac{d}{dz}\left(-\pi \cot \pi z \operatorname{cosec} \pi z\right)$$

$$= \frac{\pi}{16} \lim_{z \to -1/2} \left(\pi^2 \operatorname{cosec} \pi z \left(\operatorname{cosec}^2 \pi z + \cot^2 \pi z\right)\right)$$

$$= \frac{\pi}{16}\pi^2(-1)\left((-1)^2 + 0\right)$$

$$= -\frac{\pi^3}{16}.$$

Hence, by the second summation theorem we have

$$\sum_{n=-\infty}^{\infty} \frac{(-1)^n}{(2n+1)^3} = \frac{\pi^3}{16}.$$

The terms in this series for n and $-(n+1)$ are the same and hence

$$\sum_{n=-0}^{\infty} \frac{(-1)^n}{(2n+1)^3} = \frac{\pi^3}{32}.$$

Binomial coefficient series

In this section we see how to calculate certain series of the form

$$\sum_{n=0}^{\infty} \binom{kn}{n} x^n$$

where $x \in \mathbb{R}$.

Example 21.13

We shall show that

$$\sum_{n=0}^{\infty} \binom{3n}{n} \left(\frac{3}{64}\right)^n = \frac{2}{35}\left(7 + 3\sqrt{21}\right).$$

We have, for the contour C that gives the unit circle centred at the origin,

$$\sum_{n=0}^{\infty} \binom{3n}{n} \left(\frac{3}{64}\right)^n = \sum_{n=0}^{\infty} \left(\frac{1}{2\pi i} \int_C \frac{(z+1)^3}{z^{n+1}}\, dz\right) \left(\frac{3}{64}\right)^n, \quad \text{by Exercise 5.18(viii)},$$

$$= \frac{1}{2\pi i} \sum_{n=0}^{\infty} \int_C \left(\frac{3(z+1)^3}{64z}\right)^n \frac{1}{z}\, dz.$$

We now wish to exchange the $\int$ and $\sum$ signs so we need to use the Weierstrass M-test.

Let $a_n = \left(\dfrac{3(z+1)^3}{64z}\right)^n \dfrac{1}{z}$. Then, for $|z| = 1$, we have

$$|a_n| = \left|\left(\frac{3(z+1)^3}{64z}\right)^n \frac{1}{z}\right| \leq \left(\frac{3(|z|^3 + 3|z|^2 + |z| + 1)}{64|z|}\right)^n \frac{1}{|z|} = \left(\frac{3}{8}\right)^n.$$

As $3/8 < 1$ the series $\sum (3/8)^n$ converges, so $\sum |a_n|$ converges by the comparison test. Hence, $\sum a_n$ is convergent. From the Weierstrass M-test we can conclude that we can exchange the $\int$ and $\sum$ symbols.

Therefore, we have

$$\sum_{n=0}^{\infty} \int_C \left(\frac{3(z+1)^3}{64z}\right)^n \frac{1}{z}\, dz = \int_C \sum_{n=0}^{\infty} \left(\frac{3(z+1)^3}{64z}\right)^n \frac{1}{z}\, dz$$

$$= \int_C \frac{1}{1 - \dfrac{3(z+1)^3}{64z}} \cdot \frac{1}{z}\, dz$$

$$= \int_C \frac{64}{64z - 3(z+1)^3}\, dz.$$

We can now apply the calculus of residues. First, we find the zeros $64z - 3(z+1)^3$. As we have $(3+1)^3 - 3(z+1)^3$ it is clear that $z = 3$ is a root. After a bit more calculation involving factorization we find that the other roots are

$$z = -3 - 2\sqrt{\frac{7}{3}} \approx -6.0051,$$

$$z = -3 + 2\sqrt{\frac{7}{3}} \approx 0.0551.$$

Hence we need to calculate the residue of $64/64z - 3(z+1)^3$ at $z = -3 + 2\sqrt{7/3}$:

$$\mathrm{res}\left(\frac{64}{64z - 3(z+1)^3}, -3 + 2\sqrt{\frac{7}{3}}\right) = \frac{64}{(64z - 3(z+1)^3)'}\bigg|_{z=-3+2\sqrt{7/3}}$$

$$= \frac{64}{64 - 9(z+1)^2}\bigg|_{z=-3+2\sqrt{7/3}}$$

$$= \frac{2}{35}\left(7 + 3\sqrt{21}\right).$$

Therefore,

$$\sum_{n=0}^{\infty} \binom{3n}{n}\left(\frac{3}{64}\right)^n = \frac{1}{2\pi i}\int_C \frac{64}{64z - 3(z+1)^3}\,dz$$

$$= \frac{1}{2\pi i}\left(2\pi i \times \frac{2}{35}\left(7 + 3\sqrt{21}\right)\right)$$

$$= \frac{2}{35}\left(7 + 3\sqrt{21}\right).$$

Exercises

Exercises 21.14

(i) Show that

$$\sum_{n=1}^{\infty} \frac{1}{4n^2 + 1} = \frac{\pi \coth(\pi/2)}{4} - \frac{1}{2}.$$

(ii) Find

$$\sum_{n=-\infty}^{\infty} \frac{1}{(4n+1)(4n+3)}.$$

(iii) Find
$$\sum_{n=1}^{\infty} \frac{\coth(\pi n)}{n^5}.$$

(iv) Show that
$$\sum_{n=-\infty}^{\infty} \frac{1}{(z+n)^2} = (\pi \operatorname{cosec} \pi z)^2$$

for all $z \in \mathbb{C}\backslash\mathbb{Z}$. Hence express $\operatorname{cosec}^2 z$ as a series. Find
$$\sum_{n=-\infty}^{\infty} \frac{(-1)^n}{(z+n)^2}$$

(v) Find
$$\sum_{n=1}^{\infty} \frac{1}{n^{2k}} \text{ for } k \text{ an integer with } k \geq 2.$$

(Though it is easy to find the power of π in this one, the coefficient is very hard to find in general. For help find out about Bernoulli numbers.)

(vi) Generalize Example 21.9.

(vii) Find
$$\sum_{n=-\infty}^{\infty} \frac{1}{(2n+1)(3n+1)}.$$

(viii) Show that
$$\sum_{n=-\infty}^{\infty} \frac{(-1)^n}{n^2+1} = \frac{\pi}{\sinh \pi}.$$

(ix) Complete the details of the proof of the second summation theorem.

(x) Show that
$$\sum_{n=1}^{\infty} \frac{(-1)^{n+1}}{n^4} = \frac{7\pi^2}{720}.$$

(xi) Why can't we use the second summation theorem to find a value for the alternating harmonic series:
$$\sum_{n=1}^{\infty} \frac{(-1)^n}{n}.$$

(xii) Show that

(a) $\displaystyle\sum_{n=0}^{\infty} \binom{2n}{n} \left(\frac{4}{23}\right)^n = 2\sqrt{\frac{23}{7}}$,

(b) $\displaystyle\sum_{n=0}^{\infty} \binom{3n}{n} \left(\frac{4}{125}\right)^n = \frac{5}{28}\left(2 + 3\sqrt{2}\right)$.

(xiii) Let $a > 1$.

(a) Show that $(a+1)^3 z - a(z+1)^3$ has one root within $|z| < 1$ and two for $|z| > 1$. Find an expression in terms of a for the root within $|z| < 1$.

(b) Show that

$$\frac{1}{2\pi i} \int_{C_1} \frac{(a+1)^3}{(a+1)^2 z - a(z+1)^3}\, dz = \frac{2(a+1)}{3\sqrt{a(a+1)} - a - 4}.$$

(c) Hence find a formula for

$$\sum_{n=0}^{\infty} \binom{3n}{n} \left(\frac{a}{(a+1)^3}\right)^n.$$

Summary

❑ Suppose that

(i) $f : \mathbb{C} \to \mathbb{C}$ is analytic except for poles at $p_1, \ldots, p_m$,

(ii) $f(n) \neq 0$ for $n \in \mathbb{Z}$, and $n \neq p_j$ for any j,

(iii) there exists constants K and R such that

$$|f(z)| \leq \frac{K}{|z|^2} \text{ for all } |z| \geq R.$$

Then, the following series converge to the given expression

$$\sum_{\substack{n=-\infty \\ n \neq p_j}}^{\infty} f(n) = -\sum_{j=1}^{m} \operatorname{res}(\pi f(z) \cot(\pi z), p_j),$$

$$\sum_{\substack{n=-\infty \\ n \neq p_j}}^{\infty} (-1)^n f(n) = -\sum_{j=1}^{m} \operatorname{res}\left(\frac{\pi f(z)}{\sin(\pi z)}, p_j\right).$$

CHAPTER 22

Common Mistakes

In this chapter we look at common errors made by students learning Complex Analysis. As a lecturer with many years of experience of teaching the subject I have seen these examples appear again and again in examinations. I'm sure that, due to pressure, we've all written nonsense in an exam which under normal conditions we wouldn't have. Nonetheless, many of these errors occur every year and I suspect something deeper is going on.

What follows is not intended to be a criticism of my students, who, luckily for me, are generally hard-working and intelligent. Nor is it an attempt to mock or ridicule them. Instead the aim is to identify common mistakes so that they are not made in the future.

And if this chapter seems negative in tone, the next is more positive as it delves into techniques that improve understanding.

Imaginary numbers cannot be compared

The first mistake is the probably the most common: the comparison of imaginary numbers. For example, students write $z < R$ for z a complex number. This cannot be right. If z were $3 + 2i$ what does it mean for $3 + 2i$ to be less than R? What is usually intended is the *modulus* of z, i.e., $|z| < R$.

The point is, unlike real numbers, we cannot order the complex numbers. For example, which is bigger $3 + 2i$ or $1 + 4i$? This is difficult to decide! Since complex numbers can be identified with the plane ordering them is equivalent to ordering the points of the plane and clearly this can't be done – at least not in any useful or meaningful way. One last point needs to be made. Although $z < 3 + 2i$ is *incorrect*, note that expressions like $z < 4$ can be true if z is a real number.

Not realizing re^{it} gives a circle

This may not count as a mistake but is such a common gap in student exam knowledge that it needs to be mentioned: Too many students can't sketch the image of e^{it} and/or can't write down a contour whose image is the circle.

Let's reiterate the basics. The contour defined by $\gamma(t) = re^{it}$ defines part of a circle of radius r. A full circle can be given by $0 \le t \le 2\pi$. Furthermore, the circle can be centred at w just by adding w. That is, $\gamma(t) = w + re^{it}$ with $0 \le t \le 2\pi$ produces a circle of radius r based at w.

By taking the modulus, we can show that $\gamma(t) = re^{it}$ really does give a circle:

$$\left|re^{it}\right| = r\left|\cos t + i\sin t\right| = r\left(\sqrt{\cos^2 t + \sin^2 t}\right) = r\sqrt{1} = r.$$

Thus, the points on the contour all have the same length, i.e., are the same distance from the origin. And, a circle is just all the points the same distance from a specified point. Therefore, the image of re^{it} gives a circle when $0 \le t \le 2\pi$.

$|e^z|$ is not equal to $e^{|z|}$

Tempting as it may be to believe, $|e^z|$ is not equal to $e^{|z|}$. These two expressions are related as we shall see. The correct equality is

$$|e^z| = e^{\text{Re}(z)}.$$

It is reasonable to know the derivation of this:

$$|e^z| = |e^{x+iy}| = |e^x||e^{iy}| = |e^x| \times 1 = |e^x| = e^x = e^{\text{Re}(z)}.$$

What is the relation to $e^{|z|}$? It's an inequality:

$$|e^z| \le e^{|z|}.$$

This follows from the equality $|e^z| = e^{\text{Re}(z)}$ since exp is a strictly increasing function and $\text{Re}(z) \le |z|$, as we can see by drawing an Argand diagram. This last part is just Pythagoras' Theorem in action!

Limits of the Standard Geometric Series

Some students have trouble remembering whether the lower limit of the standard geometric series

$$\sum x^n = \frac{1}{1-x}$$

is $n = 0$ or $n = 1$.

Here is a situation in which taking a simple case provides the answer. We don't need to look it up or provide a proof. Let $x = 0$, then we have

$$\sum_{n=0}^{\infty} x^n = 1 + 0 + 0^2 + 0^3 + \cdots = \frac{1}{1-0}$$

and

$$\sum_{n=1}^{\infty} x^n = 0 + 0^2 + 0^3 + \cdots \neq \frac{1}{1-0}$$

So the lower limit is $n = 0$.

This exemplifies a useful technique: when unsure of recalling a result try a special case.

Basic definitions

It is possible to write a whole chapter on definitions and the problems that arise. (In fact, I did. See my book *How to Think Like a Mathematician.*)

Central definitions in complex analysis include differentiation and contour integral. I regularly ask for the definition of complex differentiation and regularly most of the students fail to state it correctly. Quite often what I get is

$$\lim_{h \to 0} \frac{f(z+h) - f(z)}{h}. \tag{22.1}$$

Alternatives include $f(z) \to f(a)$ as $z \to a$ or some garbled version of the above, for example with modulus signs, $z + h$ in the denominator and so on.

So what is wrong with the expression in (22.1)? For a start there is no explanation of what f, z and h are. It is important that f is a complex function and that z is identified as the point at which we are defining differentiability. Thus, we should say something like 'The complex function f is differentiable at z if ...'. Then we should bring in (22.1), making clear that we want this limit to exist, and to exist when $h \in \mathbb{C}$ rather than $h \in \mathbb{R}$.

Thus the definition should be something like 'The complex function f is differentiable at z if

$$\lim_{h \to 0} \frac{f(z+h) - f(z)}{h}, \quad h \in \mathbb{C},$$

exists'.

Similar problems occur for the definition of contour integration. That is, what γ and f are in $\int_\gamma f(z)\,dz$ is unexplained.

This may seem pedantic. It is. Pedantry is very important in mathematics.

Confusing definitions with calculation processes

Do not confuse the *definition* of an object with a *process* by which we find that object.

A good example of this mistake was mentioned in Common Error 16.20. Poles and their multiplicities are clearly central to working with residues and so I regularly ask for the definitions in exams.

In response to 'Define what it means to be a pole at w and define the order of a pole' I receive inaccurate (and often long and rambling) descriptions of how a pole is found in certain situations. For example, 'The order of the pole is the power of the thing when the pole is zero' or 'the order of the pole is the order of the bracket'. (These get no marks.) You can see in the former that the student does have some partial knowledge of pole order but that it is dependent on how pole order is calculated, i.e., we look for the zero (in the denominator) and find in most cases the power to which $(z - w)$ is taken. However, the question asked for the definition, so the definition (in this case Definition 16.16) should be given.

As another example, consider the definition of a residue. Since the most useful method for calculating a residue is Method B, I often receive $\operatorname{res}(p/q, w) = p(w)/q'(w)$ instead of a statement involving the coefficient of $(z-w)^{-1}$. Again, the student is seeing 'residue' as something that is calculated and gives a calculation method rather than seeing 'residue' as a concept.

Misstating Theorems - Insufficient detail

Following on from the previous example, if I had asked students to state Method B, then the above answer, $\operatorname{res}(p/q, w) = p(w)/q'(w)$ would be insufficient. There is no explanation of what p, q are and what conditions are placed on them.

This is a common problem. When I ask students in my geometry class or during talks I give in schools what Pythagoras' Theorem is, I receive a prompt reply: $c^2 = a^2 + b^2$. I usually, to the initial confusion of the students, say no. There are two points, one is that a, b and c are not defined. The second is deeper. The equation $c^2 = a^2 + b^2$ is the *conclusion* of the theorem. The *assumptions* are missing. The most crucial of which of course is that we need a right-angled triangle. Students certainly know this detail but ignore its importance.

Hence, for Method B we need to say 'We have $\mathrm{res}(p/q, w) = p(w)/q'(w)$ for p, q analytic with $p(w) \neq 0$, $q(w) = 0$, $q'(w) \neq 0$'.

The most common problem in misstating theorems is to just state the conclusion, particularly if the conclusion is a formula. Hopefully, now you can see that Cauchy's Integral Formula is not just

$$\int_\gamma \frac{f(z)}{z - w}\, dz = 2\pi i\, n(\gamma, w) f(w).$$

Order of Poles

Another error with poles is to believe that the order of a pole is the order of the zero in the denominator of a rational function. That this is erroneous we can see in Example 16.17(iv). The function $(e^z - 1)/z^3$ has a zero of order 3 at 0 in the denominator but the order of the pole there is 2. In non-rigorous terms – and this really is non-rigorous – the numerator has a zero of order 1 and it 'cancels' with one of the zeros in the denominator.

We can't even say that the order of the pole is at most the order of the zero in the denominator. From Example 15.6(ii) we can deduce that $\cot(z)/z^2$ has a pole of order 3 at 0, not a pole of order 2. In this case the non-rigorous explanation is that cot has a pole of order 1 at 0 so combines with the pole of order 2 for $1/z^2$ to give a pole of order 3.

This partly arises from the non-uniqueness of the representation of a function as a quotient. Here $\cot z$ is $1/\tan z$ and so $\cot(z)/z^2$ can be written as $1/(z^2 \tan(z))$. The latter representation has a zero of order 3 in the denominator (since $\tan z$ has a zero of order 1 at 0) and is non-zero in the numerator, hence we have a pole of order 3. The point, perhaps, is to not be fooled by the way a function is written as a quotient.

Integrals

Clearly, an integral of the form $\int_{-\infty}^{\infty} f(x)\, dx$, where f is a real function, must produce a real number. The methods in this book allow us to calculate such integrals with complex analysis and there is the danger that a minor miscalculation will produce an imaginary number. Hence, any working which produces an imaginary number is wrong and should be corrected.

The radius of convergence is real

The radius of convergence is real and *never* has an imaginary part.

The ratio test is one of the best tests we have for convergence of series and it can be used to calculate the radius of convergence of power series. In most elementary analysis course where real series $\sum_{n=0}^{\infty} a_n z^n$ are studied it is common for the ratio test to be stated with $a_n > 0$. This leads to some students misapplying it in the complex case. Consider the series $\sum_{n=0}^{\infty} (3 - 4i)^n z^n$ and let $a_n = (3 - 4i)z^n$. Then,

$$\frac{a_{n+1}}{a_n} = \frac{(3 - 4i)^{n+1} z^{n+1}}{(3 - 4i)^n z^n} = (3 - 4i)z \to (3 - 4i)z \text{ as } n \to \infty.$$

So far, so good. For this example, the common mistake is to write something like, 'we require that $z < \dfrac{1}{3 - 4i}$ and so the radius of convergence is $\dfrac{1}{3 - 4i}$'. (My guess is that students are slavishly following the procedure in the real case.) This cannot be right. What does a circle of radius $1/(3 - 4i)$ look like? We need the radius to be real.

To prevent this error we could define $a_n = |(3 - 4i)z^n|$ or, what amounts to the same thing, just write

$$\left| \frac{a_{n+1}}{a_n} \right| = \left| \frac{(3 - 4i)^{n+1} z^{n+1}}{(3 - 4i)^n z^n} \right| = |(3 - 4i)| \, |z| = 5|z|.$$

Thus we require $5|z| < 1$, i.e., $|z| < 1/5$. In other words the radius of convergence is $1/5$.

$\cos z$ **is not** $(z + z^{-1})/2$

I often see $\dfrac{z + z^{-1}}{2}$ substituted for $\cos z$ (and a similar substitution for $\sin z$) However, these are not equal. The confusion here comes from Chapter 20 where $\dfrac{z + z^{-1}}{2}$ is legitimately substituted for $\cos \theta$. The important differences are (i) θ is used, and (ii) it can only be used because $z = e^{i\theta}$.

Argument problems

The argument of a complex number causes a number of problems.

First, for the argument θ of $x+iy$ with $x \neq 0$ we have the equation $\tan \theta = y/x$. This means many students calculate θ with $\arctan(y/x)$ (sometimes also written, somewhat erroneously, as $\tan^{-1}(y/x)$). We can see that this can lead to errors:

$$\operatorname{Arg}(1+i) = \frac{\pi}{4} \quad \text{and} \quad \operatorname{Arg}(-1-i) = -\frac{3\pi}{4}$$

but $\arctan(1/1) = \arctan(1) = \arctan(-1/-1)$. Further details can be found in Common Error 1.7. The important point is to take care with the cases $x < 0$ and $x = 0$. Plotting on an Argand diagram often helps.

A second mistake is in the use of polar notation: If $r_1 e^{i\theta_1} = r_2 e^{i\theta_2}$, then $r_1 = r_2$ and $\theta_1 = \theta_2 + 2k\pi$ for some $k \in \mathbb{Z}$. The mistake is to forget the extra $2k\pi$ (and sometimes students take $k \in \mathbb{N}$ rather than $\mathbb{Z}$).

This leads to another common error as the previous remark has an important consequence for solving $e^z = w$. We have $z = \ln|w| + i \arg(w) + 2k\pi i$, for $k \in \mathbb{Z}$. Too often the $2k\pi i$ term is forgotten.

Odds and ends

(i) The modulus of a number is never complex. (This usually occurs due to erroneously taking $|x + iy| = \sqrt{x^2 + (iy)^2}$! This is a very, very common mistake. It leads to $\sqrt{x^2 - y^2}$ which can in turn lead to an imaginary number. See Common Error 1.6.)

(ii) $f(a + ib)$ is **not** equal to $f(a) + if(b)$. This error occurs more often than I find comfortable.

(iii) If z is complex, then $z \to \infty$ has not been defined. However, $|z| \to \infty$ is ok.

(iv) If $f(x + iy) = u(x, y) + iv(x, y)$, then v_x is **not** $\dfrac{\partial}{\partial x}(iv)$.

(v) The definition of contour integral does not include a modulus.

(vi) Cauchy's Theorem requires that the path γ is *closed*.

Summary

- ❏ Complex numbers cannot be compared with $<$ and $>$ signs.

- ❏ $w + re^{it}$ gives part or all of a circle.

- ❏ $|e^z| \leq e^{|z|}$ and is not an equality in general but $|e^z| = e^{\mathrm{Re}(z)}$.

- ❏ The lower limit of $\sum x^n = \dfrac{1}{1-x}$ is $n = 0$ not $n = 1$.

- ❏ Reproduce basic definitions verbatim.

- ❏ Don't confuse definitions with calculation processes.

- ❏ State theorems correctly – include the assumption, don't just give the conclusion, particularly for formulae.

- ❏ The order of a pole is not 'the order of the zero in the denominator'.

- ❏ Real integrals do not evaluate to give a complex number.

- ❏ The radius of convergence is real, it **never** has an imaginary part. Take the modulus when working out $\dfrac{a_{n+1}}{a_n}$ for radius of convergence.

- ❏ $\cos z$ is **not** equal to $\dfrac{z + \frac{1}{z}}{2}$. Similarly for sin.

- ❏ The argument of a complex number can't always be calculated by just $\tan^{-1}(y/x)$.

- ❏ Don't forget that the angle θ and $\theta + 2k\pi$ are the same for $k \in \mathbb{Z}$ and so in many formulae there is an extra term to deal with this ambiguity.

CHAPTER 23

Improving Understanding

A course is more than a collection of definitions, theorems and proofs. If the teacher has done their job correctly, then there is a common theme and a well-defined stock of techniques and ideas. For true understanding of mathematics one should look for the theme and for the repeatedly used techniques.

Complex analysis is no exception. The theme I have chosen in this book is to generalize the notions of calculus to the complex numbers in order to solve problems involving real numbers. The chapters on real integrals are good examples of this. The statements of the problems are purely real – e.g., sum a real series – while the solution is found by going through the complex domain. It is possible to create a course that omits these applications and generalizes for the sake of generalization and the motivation for study comes from the elegance of the resulting theory.

The main technique used here is the Estimation Lemma. However, the main *theoretical* result upon which I build results is Cauchy's Theorem. From it we prove Cauchy's Integral Formula, Taylor's Theorem and Cauchy's Residue Theorem. In each of these the Estimation Lemma was used but it is Cauchy's Theorem that makes Complex Analysis what it is.

In any course, just as important as to what to include is what to exclude, or at least not emphasize as much. One area I have underplayed is topology. I could have included some more point set topology theorems. These could have been used in the proof of the Identity Theorem for example. Instead I opted to build into proofs any required topology results. Another topological theorem is the Paving Lemma. This is considered central in the classic textbook *Complex Analysis* by Stewart and Tall but here it is relegated to Appendix A to help prove the general form of Cauchy's Theorem.

Basics of improving understanding

Higher level mathematics involves a stronger emphasis on statements and proofs rather than simple techniques such as finding the roots of a quadratic. In Complex Analysis we have a good blend of theory and simple techniques. For example, Cauchy's Residue Theorem relies on some serious theoretical results but its application involves finding winding numbers and residues, both of these usually involves simple calculations.

For the theoretical results it is very important to state the theorems with care. As described in the Common Mistakes chapter we do not state a theorem by just giving the conclusion, for example stating Cauchy's Integral Formula theorem as only

$$\int_\gamma \frac{f(z)}{z-w}\, dz = 2\pi i\, n(\gamma, w) f(w).$$

It is crucial that the assumptions are included and for deeper understanding you need to know *why* they are included. If an assumption is dropped or changed, then what difference does it have on the result?

The assumptions can be crucial in subtle ways. This can be seen in Theorem 8.9, The Fundamental Theorem of Calculus. One of the conditions is that there exists F such that $F' = f$. It does not say that there exists one. Not realizing this and internalizing it leads to the common misconception that one could directly prove Cauchy's Theorem from it. (The fallacious argument is that $\int f(z)\, dz = F(\gamma(b)) - F(\gamma(b))$ and since the contour is closed, $\gamma(a) = \gamma(b)$ and so $\int f(z)\, dz = 0$.) Note here the use of the word 'directly'. This is again subtle. The Fundamental Theorem of Calculus *is* used in the proof of the simplified version of Cauchy's Theorem given earlier and in the full version in the Appendix! The point is that it is merely a small part of each proof.

One can get a further understanding of statements by consulting other books as this will often give insight into the essential part of a theorem. For example, in Stewart and Tall (see Further Reading below), Cauchy's Integral Formula is only stated for a circular contour traversed once. Priestley's statement (again, see Further Reading below) is only for simple closed positively oriented contours that are traversed once. If you look up her statement, then be aware that, unlike in this book, Priestley defines contours to be simple and closed.

Some common patterns

The power of the Estimation Lemma has already been remarked upon. What other ideas are used repeatedly? One is the paradigm introduced in Remark 2.15. That is we prove results in complex analysis from results in real analysis. This is achieved by taking the modulus and the argument (or just one of them), or by taking real and imaginary parts. In fact, this can be seen throughout the book since the Estimation Lemma is a central example of this idea.

It can also be seen in results such as proving that complex power series are differentiable and differentiable term-by-term, the ratio and comparison tests used there are both real analysis results.

A lot of results rely on taking the modulus so here's a quick useful tip. Instead of using $|z|$, use $|z|^2$. This eliminates the square root so reduces clutter in the calculations and also we can clearly use $z\bar{z}$ as a possible substitution to help solve the problem.

The Estimation Lemma

As stated earlier, the Estimation Lemma appears regularly and so to gain deeper understanding of Complex Analysis through learning and understanding proofs we should ask 'Where does the Estimation Lemma appear?' The proof can then be broken into parts: 1) getting to the Estimation Lemma, 2) the consequence of using the lemma. The next section deals with this more generally.

Memorizing proofs

Reproducing proofs is important for exams and many students attempt to memorize proofs word-for-word. This is very inefficient. Proofs are not like lines in a play that need to be spoken precisely as given. In a particular proof there is usually some structure that we need to know and we can just improvise the rest around it. Learning proofs this way leads to deeper understanding of the material and moreover develops mathematical abilities.

In the previous section I claimed that finding where the Estimation Lemma is used reveals the structure of many proofs in Complex Analysis. If you haven't already done so, go through and identify proofs which use it. (You should see most do!) Thus when memorizing proofs in Complex Analysis one should aim to memorize where in the proof the Estimation Lemma is used. To apply the

Estimation Lemma we need a function and a contour. Those are the particulars we have to be memorize.

Let us take the proof of Liouville's Theorem as an example. The key point is to apply the Estimation Lemma to the integral in Taylor's Theorem that allows us to calculate the first derivative of the function. The contour is just a circle at z_0 and like in many other proofs and examples we let the radius go to infinity.

I would remember the proof simply as 'Use the Estimation Lemma on the integral from Taylor's Theorem to find the first derivative at a general point'.

Note that I don't write down extras such as 'Show $f'(z_0) = 0$ by letting the radius tend to infinity'. The reason for this is that most of the proofs and examples have that the radius goes to infinity or to zero. After all, what else are we going to do with it? It should be obvious which we should in each application. Here letting r go to zero would make the bound go to infinity which is useless. Thus we let r go to infinity. From there it is obvious that $f'(z_0) = 0$. It should then be equally obvious that this means f is constant since z_0 was general.

To recap, we memorize this proof in one line. For other proofs we may have to use more steps. For example, in the proof of the Fundamental Theorem of Algebra, I would use

(i) Proof by contradiction.

(ii) Show bound $\left|\dfrac{p(z)}{z^n}\right| \geq \dfrac{|a_n|}{2}$ for large $|z|$.

(iii) Apply Estimation Lemma to Cauchy's Integral Formula for calculating $1/p(0)$.

(iv) Show $1/p(0)$ is 0 for the contradiction.

The proof is then as follows: From step (i) I know that my polynomial p has no roots. Step (ii) reminds me to define a_j in the polynomial and to work out how $\left|\dfrac{p(z)}{z^n}\right|$ behaves for large $|z|$.

Note that if I do not succeed in proving the correct bound for some reason (such as exam nerves), then by memorizing the bound I need, I can at least continue the proof from this point.

Next we apply the Estimation Lemma to an integral arising from Cauchy's Integral Formula. This gives $|1/p(0)| = 0$ which is impossible and so we have the required contradiction.

Knowing facts versus looking them up

It is currently popular to claim that students should not learn anything they could look up. Inspirational quotes by Einstein such as 'Education is not the learning of facts, but the training of the mind to think' and 'Never memorize something you can look up' are shared on social media. Leaving aside whether he said these or not, the sharer's point is fairly clear, memorization of facts has no place in a respectable education.

It is equally clear where this idea comes from. In the past much education was about reciting facts – I certainly spent a lot of my early mathematics lessons chanting times tables along with my classmates. Such memorization without understanding is dull and frustrating for the learner and it is easy to have sympathy with the 'look it up approach'. Too much memorization of facts and too little understanding is uninteresting.

Despite this, I'm going to make a case for knowing facts that could be looked up.

A student came to my office with a complex analysis problem he was struggling with. My reaction in this type of situation is not to explain the answer on the board but to hand the student the pen and make them write whilst I ask probing questions in my, admittedly poor, impersonation of Socrates. My reasoning here is that I know that if I only explain, the student will nod, say yes to anything I say and will write down the answer in the hope of understanding it later. Having the pen and writing forces them to think and clarifies for me where their misunderstanding lies. Yes, it is more painful for them (and me) but they learn a lot more.

This particular student was progressing well in the solution until we got stuck because he needed the sine addition formula and didn't know it by heart. (The formulae is $\sin(A + B) = \sin A \cos B + \cos A \sin B$ and a method for deducing it is given in the next section.) He claimed that he didn't need to know it as it was something he could look up when needed. Well, clearly, this wasn't true. He needed it there and then and wasn't able to look it up. Interestingly, and bafflingly, he repeatedly refused to let me tell him how to quickly deduce it. In the end he left my office promising that he would be able to sort out the problem.

Why am I concerned about this? Well, we had been diverted by a triviality. It is akin to halting in an arithmetical calculation to look up $7 + 5$. This minor problem had broken our concentration and focus on the important question. When learning a foreign language – and mathematics is very much a language – one could not achieve fluency if one had to always look up vocabulary or how to conjugate a verb. Here lack of knowing impacted fluency. Knowing enables fluency.

But let's also look at the wider picture. What is learning? How can we define it? One definition is 'the transfer of facts and skills into long term memory'. You really know something when you can instantly remember it. For example, you can claim to have learned the basics of driving when the actions involved are pushed down from the conscious to the subconscious. No one can claim to be able to drive if, every time, they had to consciously think through how to overtake or how to change gear.

Another important part of learning is to make connections between topics and facts. We often feel we have gained a deep insight when we see how two seemingly different areas are connected. For example, viewing complex numbers as vectors in the plane. But how can you make a connection between facts if you don't know those facts?

So there are some serious reasons to memorize facts and procedures. In particular to not waste time, improve fluency and to make connections between areas.

That is not to say one should memorize every fact from every course taken and I should make clear that I am not in favour of mindless rote learning. The memorization is there to aid problem-solving and understanding and is definitely not the goal of education.

My favoured approach is to abandon the choice between the "memorization" and "look-it-up" mindsets. The key skill is to identify what should be memorized and what should be looked up. For any particular fact it becomes a matter of choice and depends on what one uses regularly. Something used day-to-day will likely become memorized anyway. It is the material that is used every so often where a decision needs to be made.

My belief is that the sine formula should be memorized as it is used in different areas – analysis, geometry, and ordinary differential equations to name a small selection. Whether something like the Cauchy-Riemman equations will need to be memorized will depend on the course of study. Obviously it should be learned by heart for an exam as generally one can't look up in an exam and besides it would waste time working it out. (A method to remember the Cauchy-Riemann equations is given on page 85.) Anyone who uses the equations regularly should know it by heart.

As another example, outside of teaching I rarely use half angle formulae (or equivalently double angle formulae), for example, $\cos t = 1 - 2\sin^2(t/2)$. However, I can deduce them in seconds from the sine and cosine addition formulae (which I do know). I know this will horrify some of my colleagues as they know them by heart but as I rarely use them I feel that I don't need to keep them in my head. It leaves space for other information!

How to remember the sine formula

In case you are interested, here's a method to remember the sine formula that was mentioned in the previous section. In essence, it is the method from Exercise 1.14(xviii):

We have two simple facts that we should certainly know without looking them up as they are used so often: (1) $e^{i\theta} = \cos\theta + i\sin\theta$ for $\theta \in \mathbb{R}$, and (2) $e^{i(A+B)} = e^{iA}e^{iB}$ for $A, B \in \mathbb{R}$. Thus we have

$$e^{i(A+B)} = e^{iA}e^{iB}$$
$$\cos(A + B) + i\sin(A + B) = (\cos A + i\sin A)(\cos B + i\sin B)$$
$$= \cos A\cos B - \sin A\sin B$$
$$+ i(\sin A\cos B + \cos A\sin B).$$

Equating real and imaginary parts gives *both* the sine and cosine addition formulae.

Sometimes, when a student half-remembers the formula but is unsure whether it is plus or minus or whether it involves sine times cosine or sine times sine, then I ask them to write their guess and then use $B = 0$ and $B = \pi$ to check it. In most guesses, doing that the correct formula becomes clear.

Expanding power series and calculating residues

The preceding sections are rather abstract and so I will finish on an example of a specific technique that can save time. The broader abstract principle to learn is that one should always look for places where calculations can be simplified. Lecturers often do not give the most efficient way of doing something. There are various reasons for this approach, the most efficient way may lack any motivation or it may involve learning too much in one go. Hence it is often useful to return to a topic and ask how the calculations could be simplified.

Method D for calculating residues in Chapter 17 can involve some tiresome calculations of Laurent series. One way to avoid tedium is to note that we only need the coefficient of z^{-1} and hence we do not need to calculate the whole series.

For example, consider $\dfrac{\sin z}{z^6(2z - 1)}$ at 0. The z^6 in the denominator means that we should find the coefficient of z^5 in the Taylor series expansion of $\dfrac{\sin z}{2z - 1}$. (Note that it is Taylor series rather than Laurent as this new function is defined and differentiable at $z = 0$.) We can find the coefficient of z^5 without calculating the whole Taylor series (or at least not all the terms up to z^5). To see how do we do this let's first see what we have to do to calculate the Taylor series.

We can multiply the Taylor series of $\sin z$ with that of $1/(2z - 1)$. These series are

$$\sin z = z - \frac{z^3}{6} + \frac{z^5}{120} + \ldots, \text{ and}$$

$$\frac{1}{2z - 1} = -\frac{1}{1 - 2z} = -\sum_{n=0}^{\infty}(2z)^n = -1 - 2z - 4z^2 - 8z^3 - 16z^4 - 32z^5 + \ldots$$

Multiplying these out is complicated, time-consuming and it is easy to make a small error. However, we need only the z^5 terms and so require the sum of the coefficients arising from multiplying the constant and z^5 terms in the respective series, the z and z^4 terms and so on.

This is easily done by drawing a quick table:

		z^0	z^1	z^2	z^3	z^4	z^5
		0	1	0	$-\frac{1}{6}$	0	$\frac{1}{120}$
z^0	-1						$-\frac{1}{120}$
z^1	-2					0	
z^2	-4				$\frac{2}{3}$		
z^3	-8			0			
z^4	-16		-16				
z^5	-32	0					

The z^5 coefficient is the sum of the terms on the (anti-)diagonal, in this case

$$-16 + \frac{2}{3} - \frac{1}{120} = -\frac{1841}{120}.$$

Hence, we can now find the residue we seek:

$$\operatorname{res}\left(\frac{\sin z}{z^6(2z - 1)}, 0\right) = -\frac{1841}{120}.$$

Once this method is understood one can streamline it further by not drawing a table but just writing down the relevant products.

Further Reading

If you are interested in further reading, then the following are of interest.

❑ Complex Analysis, I. Stewart and D. Tall. This is a classic and includes more on homotopies and is good for comparison of their statements with ours.

❑ Introduction to Complex Analysis, H.A. Priestley. Another classic, more concise than this book and Stewart and Tall. Two good chapters for further reading are the one on Laplace and Fourier Transforms and the one on conformal mappings. The latter is a good introduction to the geometry of complex functions.

❑ The Riemann Hypothesis is a statement about what is called the Riemann Zeta function. The hypothesis is that the non-trivial zeros of the function lie on the line of complex numbers with real part equal to $1/2$.

This is a long-standing conjecture that currently has a large prize associated with it. Prove or disprove the Hypothesis and you will win a million dollars.

The function has a strong connection to prime numbers and so is important in Number Theory.

Good books in this area are *The theory of the Riemann zeta-function*, E.C. Titchmarsh, OUP, (1986), and *The Riemann Hypothesis, A Resource for the Afficionado and Virtuoso Alike*, ed, P. Borwein, S. Choi, B. Rooney, A. Weirathmueller, Springer, (2008).

Summary

❑ Assumptions are an important part of theorems. It's not all formulae.

❑ The Estimation Lemma is a key tool in Complex Analysis.

❑ Take the modulus and move from complex to real.

❑ Memorize proofs not line by line but by structure.

❑ Some facts should be known. Looking them up can't always be done, wastes time, interrupts flow and prevents connections being made.

❑ Expanding power series is prone to errors so only focus on the terms you need.

APPENDIX A

Proof of Cauchy's Theorem

We shall now prove the version of Cauchy's theorem given in Chapter 11.

Theorem A.1
Let $D \subseteq \mathbb{C}$ be an open set, and $f : D \to \mathbb{C}$ be a differentiable complex function. Let γ be a closed contour such that γ and its interior points are in D.

Then, $\displaystyle\int_\gamma f = 0$.

Once Cauchy's Theorem has been proved for a finite sum of straight line contours and arcs we can use Exercise 11.12(ii) on the existence of derivatives to give a proof of the full version. First we need a lemma.

Lemma A.2 (Paving Lemma)
Let D be an open set in $\mathbb{C}$ and $\gamma : [a, b] \to D$ be a contour. Then there exists $n \in \mathbb{N}$, $t_j \in [a, b]$ with $j = 0, 1, 2, \ldots, n$, and open discs $D_j \subseteq D$ centred at $\gamma(t_j)$, $j = 0, 1, 2, \ldots, n - 1$ such that

(i) $a = t_0 < t_1 < t_2 < \cdots < t_n = b$,

(ii) $\gamma([t_j, t_{j+1}]) \subset D_j$,

(iii) $\displaystyle\bigcup_{j=0}^{n-1} D_j \subseteq D$.

That is, we can divide the contour into pieces so that the image of each piece lies in an open disc and the union of these discs lies in D.

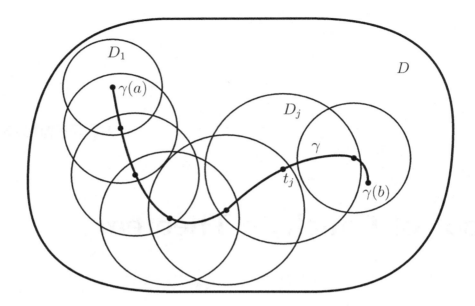

Proof. For any $c \in [a, b)$ there exists an open disc D' around $\gamma(c)$ with $D' \subseteq D$. From the continuity of γ there exists $\delta > 0$ such that $\gamma([c, c+\delta)) \subset D'$ and hence $\gamma([c, c+\delta/2]) \subset D'$.

Now let $\Lambda \subseteq [a, b]$ be the set of points $t \in [a, b]$ such that the conclusion of the Paving Lemma holds for $\gamma|[a, t]$. Then, Λ is non-empty since by the above with $c = a$ we have that $a + \delta/2 \in \Lambda$. As Λ is a non-empty set of real numbers and is bounded above by b, Λ has a least upper bound, l say, by the completeness axiom.

Assume $l < b$. Then taking $c = l$ in the above we see that $l + \delta/2 \in \Lambda$ contradicting the least upper bound property of l. Hence $l = b$. $\qquad\square$

Proof (of Cauchy's Theorem). From the Paving Lemma there exist $t_j \in [a, b]$ and open discs D_j.

Let $\widetilde{\gamma}_j$ be the straight line contour from $\gamma(t_{j-1})$ to $\gamma(t_j)$. As these points lie in D_{j-1} then $\widetilde{\gamma}_j^* \subseteq D_{j-1}$. Let $\widetilde{\gamma} = \sum_{j=1}^{n} \widetilde{\gamma}_j$. Then, as $\cup_{j=0}^{n-1} D_j \subseteq D$ we can prove that $\text{Int}(\widetilde{\gamma}) \cup \widetilde{\gamma}^* \subset D$. Hence, as $\widetilde{\gamma}$ is a finite union of straight line contours we know from the simple version of Cauchy's Theorem that $\int_{\widetilde{\gamma}} f = 0$.

Let $\gamma_j = \gamma|[t_{j-1}, t_j]$, $j = 1, \ldots, n$. Then $\gamma = \sum_{j=1}^{n} \gamma_j$ and $\gamma_j^* \subset D_{j-1}$. By Exercise 11.12(ii) the function $f|D_{j-1}$ has an antiderivative, F say. Thus by the Fundamental Theorem of Calculus,

$$\int_{\gamma_j} f = F(\gamma(t_j)) - F(\gamma(t_{j-1})) = \int_{\widetilde{\gamma}_j} f.$$

Finally we have,

$$\int_\gamma f = \sum_{j=1}^n \int_{\gamma_j} f = \sum_{j=1}^n \int_{\tilde{\gamma}_j} f = \int_{\tilde{\gamma}} f = 0.$$

Well, that seems a fun place to end the book! $\qquad \square$

Index

Made in the USA
Las Vegas, NV
19 November 2023

81134769R00155